THE VOLTERRA SERIES AND ITS APPLICATION

BY

MARK ROBERT DUNN

B.E.E (Georgia Institute of Technology) 1984

THESIS

Submitted in partial satisfaction of the requirements
for the degree of

MASTER OF SCIENCE

in

ELECTRICAL AND COMPUTER ENGINEERING

in the

GRADUATE DIVISION

of the

UNIVERSITY OF CALIFORNIA

DAVIS

Approved:

Richard Spencer 8/27/92

Stephen H. Lewis 8-27-92

W. A. Gardner 8/28/92

Committee in Charge

1992

ISBN-13: 978-1508669906

ABSTRACT

Modeling of weakly nonlinear systems by means of Volterra series analysis is presented. Necessary conditions for representing nonlinearities by a Volterra series are developed analytically as well as heuristically. A two-condition convergence criterion for Volterra series and a method for determining Volterra transfer functions are established.

For systems with multiple nodes, an extension of Volterra series analysis; method of nonlinear currents is developed and applied to a MESFET amplifier. Finally, methods of quantifying nonlinear behavior are discussed.

ACKNOWLEDGMENT

I would like to thank Hewlett Packard for providing the financial support through its Fellowship program. I am especially grateful to Kent Stalsberg. I owe my acceptance in the Fellowship program to his perseverance, tenacity and ultimately to his belief in me. I also thank Richard Spencer for his flexibility, patience and advice.

Finally, and most importantly, I thank my wife Sue without whom none of this would have been possible.

Table of Contents

TABLE OF FIGURES

1.0 INTRODUCTION

Many papers and books have been devoted to the analysis and synthesis of linear systems, in spite of the fact that no system is entirely linear. All systems, whether electrical, mechanical, biological, or whatever, exhibit varying degrees of nonlinear behavior.

In medicine a patient's condition may improve with a single dose of medication, be cured with a double dose or made critical with a triple. In electrical engineering one is constantly faced with nonlinearities due to the fact that the constitutive relationships of any electrical device are nonlinear if a large enough range is considered. Diodes, for example, are nonlinear regardless of the input, whereas resistors will be nonlinear only at the extremes of their operating range, when the power dissipated causes thermal effects to vary their resistance.

Generally when an engineer is faced with a nonlinear problem one of two approaches is taken. The first approach is to set up a measurement system that quantifies the problem (i.e. measures the power of an undesired harmonic) and then proceed to tweak the design until the undesired response is minimized. This empirical approach sometimes

9

works but is unsatisfying to most engineers. The second approach is to try to develop a better theoretical understanding of the problem by applying linear system theory to a nonlinear phenomenon. Both methods lead to frustration due to the fundamental lack of understanding of the problem.

Techniques for analyzing and designing nonlinear circuits do exist, although they are not well represented in many textbooks. The purpose of this thesis is to demystify one of these techniques, Volterra series, by providing both an intuitive and mathematical basis for its existence. In this thesis I have pulled together many diverse treatments of Volterra series analysis and merged them into a unified, cohesive pedagogical tool. By drawing from a variety of disciplines and approaches, my goal is to develop a well-rounded, complete treatment that not only conveys a mathematical understanding of a Volterra series but also an intuitive or engineering sense of its strengths, limitations and ultimately its applicability to modern nonlinear networks.

1.1 DEFINITION OF NONLINEAR SYSTEMS

A system is said to be nonlinear if it doesn't obey superposition. Mathematically this can be stated as follows:

If $Y_1 = T[X_1]$ and $Y_2 = T[X_2]$

where T is the transfer function for an arbitrary system, X is the input and Y is the output

then for the system to be linear:

$c_1 Y_1 + c_2 Y_2 = T[c_1 X_1] + T[c_2 X_2]$

must be true.

An important implication of this definition is that there will be no frequency found at the output of a linear system that was not applied at the input. Conversely, the output of a nonlinear system will contain frequencies that are related to, but not necessarily equal to, the frequencies of the input signals. The number of new frequencies and their amplitude will depend on both the amplitude of the input signal and the severity of the nonlinearity (i.e. how many terms of a power series are required to accurately represent the nonlinearity).

1.2 STRONGLY vs. WEAKLY NONLINEAR SYSTEMS

Mathematically stated, describing a system as weakly nonlinear requires that its full range of inputs lies within the radius of convergence of a power series expansion of that system. This further means that both the constitutive relationship modeled by the power series and its derivatives are continuous. This definition comes from the fact that a

Volterra series is a generalization of a Taylor series expansion of a function.

From a practical point of view, weakly nonlinear implies that the nonlinear system can be accurately represented by a power series expansion of only a few terms. When more than three terms are required the algebraic complexity of calculating higher-order Volterra coefficients becomes unwieldy. For strongly nonlinear circuits, such as samplers, switches or any other system with discontinuous constitutive relationships, or systems with large inputs such as mixers, other techniques such as harmonic balance need to be employed. In this thesis it will be assumed all systems are weakly nonlinear.

1.3 METHODS OF ANALYZING NONLINEAR SYSTEMS

1.3.1 PIECEWISE LINEAR [26]

In this method all nonlinearities are represented by piecewise linear approximations. The obvious appeal of this technique is that linear system theory can be applied. A disadvantage is that this method is computationally intensive and provides very little insight into the nonlinear phenomenon.

1.3.2 NONLINEAR DIFFERENTIAL EQUATIONS [26]

Applying physical laws, any nonlinear system can be described by a set of nonlinear differential equations. For lumped electrical circuits these equations can be formulated using conventional circuit theory. This method has many drawbacks. The Laplace transform, which reduces linear differential equations to a set of algebraic equations, is not valid for nonlinear equations. Uniqueness of the solution is also not guaranteed as it is for linear systems.

Another problem exists in specifying the initial conditions of the system. For some initial conditions the response of a nonlinear system may never reach a steady-state. Such systems are called chaotic. For other initial conditions the response may behave chaotically before reaching a periodic steady-state. Enough time must be allowed to elapse before declaring a system chaotic. This leads to the natural question of how much time is "enough".

Because of these questions and concerns this method is usually not preferred for nonlinear analysis.

1.3.3 HARMONIC BALANCE [8]

Modern harmonic balance is basically a hybrid of frequency- and time-domain techniques. A nonlinear system can be represented by a combination of linear and nonlinear subnetworks. Linear subnetworks are analyzed in the frequency domain while the nonlinear subnetworks are analyzed in the time domain. The analysis procedure can be described as follows:

1. Separate the nonlinear network into linear and nonlinear subnetworks. Refer to Figure 1.1.
2. Decide upon the number of significant harmonics, N, to be contained in the solution. A small value of N minimizes computational time, however, too small a value leads to aliasing errors in later steps.

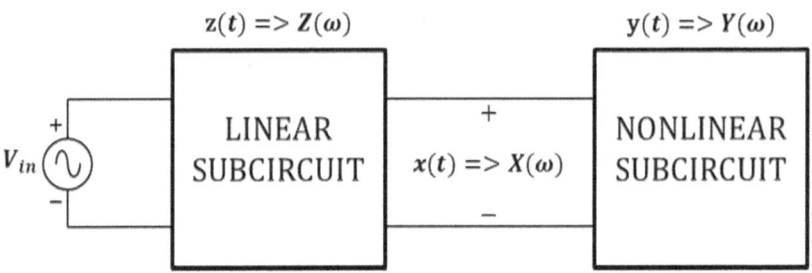

Figure 1.1 Block Diagram of Typical Nonlinear System

3. Impose a signal containing N harmonics, $x(t)$, at the interface between the linear and nonlinear subnetworks. In this example, $x(t)$ can represent a voltage and $y(t)$ and $z(t)$ the currents entering or leaving the nonlinear and linear subcircuits respectively. The initial amplitude of the signal is guessed. One method for estimating an initial

amplitude is to use the solution of the linearized approximation of the nonlinear network. This is achieved by representing the nonlinear elements by only their first-order terms.

4. Calculate the time response, $y(t)$, of the nonlinear network to this input.
5. Perform a Fourier transform on $x(t)$ and $y(t)$.
6. Determine the response, $Z(w)$, of the linear subnetwork to the input $X(w)$
7. Evaluate the difference $\epsilon = Z(w) - Y(w)$
8. Optimize $x(t)$ until ϵ is below a predefined threshold.

This method is more efficient than a purely time-domain approach since it takes advantage of the relative ease of analyzing the linear portion of the system in the frequency domain. Another advantage is that it works for strongly nonlinear networks by choosing larger numbers of harmonics to use in the analysis. This method obviously is computationally intensive and therefore, requires a computer. In order to solve nonlinear systems in a reasonable time frame, efficient Fourier and inverse Fourier transform techniques must be employed.

Inherent in this method for solving nonlinear systems are the standard optimization concerns. For a solver to be reliable the solutions must converge for a stable system. The error function must also be able to converge to the global rather than a local minimum.

Very little insight into the system is gained because of the computational complexity of this method and the fact that once the harmonic number, N has been set, the computer takes over.

1.3.4 VOLTERRA SERIES [23]

A fourth approach, and the one that this thesis will deal with, uses functional series to represent nonlinear systems. Volterra first studied the functional series named after him in the 1880's as a generalization of the Taylor series expansion of a function [27]. Fréchet later showed that an infinite series of functionals can represent the input-output relationship of all continuous nonlinear systems.

Some of the advantages of Volterra series include:

- Explicit input-output relationships of nonlinear systems.
- Straightforward methodology for handling combinations of nonlinear systems.

An additional major advantage lies in the fact that Volterra series analysis is an extension of concepts used in linear system theory. Engineers aren't required to use a totally alien tool to handle nonlinear systems. On the downside, this method is computationally prohibitive for the higher-order terms associated with strong nonlinearities. Therefore, Volterra series expansions are usually truncated after the third term. A Volterra series also has an associated radius of convergence. This limits the range of inputs for which a Volterra series exists. For these reasons a Volterra series is only suitable for weakly nonlinear systems.

2.0 MODELING SYSTEM NONLINEARITIES

2.1 INTRODUCTION

Inherent in the following discussion on modeling and nonlinear analysis is the quasistatic assumption [11]. This assumption requires that all dependent nonlinear parameters respond instantaneously to changes in their control sources. Currently, the propagation delays in high-frequency active devices are short enough to allow a quasistatic approximation. As devices are pushed higher and higher in frequency it is clear that at some point this assumption will introduce sufficient errors requiring a more rigorous approach to modeling high-frequency devices.

The first step in modeling a nonlinear system is to quantify its behavior. This is typically achieved by measuring the constitutive relationship which defines the particular nonlinear system (i.e. I vs. V, C vs. V, Q vs. V, etc.). The next step is to somehow represent this tabular information in a more analytical or functional form. The degree of nonlinearity exhibited by a system is very difficult to determine from a table of data. For this reason a curve-fitting technique is usually employed to analytically represent the particular nonlinear system.

2.2 POLYNOMIAL REPRESENTATION

The first step in curve fitting is to determine what type of function is needed to represent the given nonlinearity. A very useful and convenient form is the polynomial:

$$f(x) = \sum_{n=1}^{N} a_n x^n \qquad (2.1)$$

where N is the order of the approximation

An important feature of polynomials is that they and their derivatives are continuous. This characteristic will be exploited later. The existence of such a representation is guaranteed by the Weierstrass theorem [21] which states:

If $f(x)$ is a continuous, real-valued function on the closed interval $[a, b]$, then given any $\epsilon > 0$ there exists a real polynomial $p(x)$ such that $|f(x) - p(x)| < \varepsilon$ *for all* $x \in [a, b]$.

The assumption that the interval $[a, b]$ is closed and bounded is essential. Outside the interval no conclusion regarding the system can be drawn. The polynomial may radically diverge from the data.

2.3 CURVE-FITTING TECHNIQUES [29]

Now that a form has been determined for the approximating function, the next step is to choose the appropriate technique for fitting the data to the function. The two basic techniques are interpolation and approximation. In interpolation the approximating polynomial is forced to pass through each known data point whereas in approximation no such constraint exists. In an approximation technique the only requirement is that the curve come "as close as possible" to each measured data point. The most common measure of "as close as possible" is the least squares criterion. The application of this criterion is known as the method of least squares.

Since interpolation polynomials fit the measured data exactly, they and any formulas based on them are seriously influenced by noise or errors in the data. On the other hand, a polynomial fitted to the data via approximation will be less influenced by random errors and will better represent the underlying, presumably smooth trend of the data. Therefore, derivatives computed from such approximating functions will in general be more accurate than those computed by differentiating interpolation polynomials. This point is illustrated below in Figure 2.1. Note that the instantaneous slope of the interpolation function fluctuates radically from point to point while the slope of the least square approximation changes in a smooth consistent manner which is almost certainly a more accurate representation of the trend the data would exhibit if free from random error.

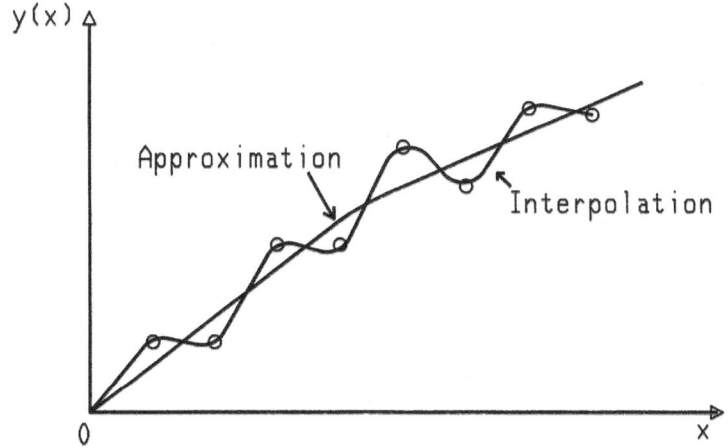

Figure 2.1 Interpolation vs. Approximation

In electrical engineering, some constitutive relationships of interest are I/V and Q/V. An element that is described by an I/V characteristic can generally be regarded as a voltage-controlled conductance. Similarly a Q/V characteristic can be represented as a voltage-controlled capacitor. Current-controlled inductors, represented by Φ/I, are rarely encountered.

Once the desired constitutive relationship has been established, the next step is to measure the nonlinear system to be modeled. For example, modeling a nonlinear conductance is generally accomplished by measuring current vs. voltage. An appropriate approximation algorithm is employed to fit the raw measured data to a polynomial. The order, N, of the approximating polynomial is determined by the maximum allowable error between data and approximation. A high value of N produces a small error, ϵ, as guaranteed by Weierstrass.

The result of the above described procedure is an Nth-order polynomial that represents the desired nonlinear element. For example, a nonlinear resistor would be modeled by a polynomial of the form:

$$I = a_1 V^1 + a_2 V^2 + a_3 V^3 + \cdots + a_N V^N \qquad (2.2)$$

$$I = \sum_{n=1}^{N} a_n v^n \qquad (2.3)$$

A common mistake is to equate a_n with the nth-order incremental conductance of the system i.e.

$$I = g_1 V^1 + g_2 V^2 + g_3 V^3 + \cdots + g_N V^N$$

$where\ g_n = a_n$ [11].

22

This assumption is only valid @$V = 0$:

$$\frac{d^n I}{dV^n}\Big|_{@V=0} = g_n = a_n \tag{2.4}$$

Generally the nonlinear characteristics of interest do not occur at $V = 0$, but rather at some bias point $V = V_0$. Whenever a nonlinear system has some bias, assuming g_n represents the incremental conductance of the nth-order nonlinearity is incorrect.

To establish the proper incremental conductance, the global I/V approximating polynomial needs to be expanded in a Taylor series around the bias point V_0.

$$I(V_0 + v) = I(V_0) + \frac{dI}{dV}\Big|_{V=V_0} \cdot v + \frac{1}{2!}\frac{d^2 I}{dV^2}\Big|_{V=V_0} \cdot v^2$$
$$+ \frac{1}{3!}\frac{d^3 I}{dV^3}\Big|_{V=V_0} \cdot v^3 \cdots \tag{2.5}$$

Where V_0 is the bias voltage and v is the small-signal input voltage. From this expansion the small-signal current, $i(t)$, can be isolated by subtracting from (2.5) the current due to the bias voltage V_0.

$$i(t) = I(V_0 + v) - I(V_0) = \frac{dI}{dV}\Big|_{V=V_0} \cdot v + \frac{1}{2!}\frac{d^2 I}{dV^2}\Big|_{V=V_0} \cdot v^2$$
$$+ \frac{1}{3!}\frac{d^3 I}{dV^3}\Big|_{V=V_0} \cdot v^3 \cdots \tag{2.6}$$

Note that in the above expression only the bias current due to V_0 was removed and not all the direct current [11]; this is a subtle yet important point. $i(t)$ represents the total small-signal current which includes both ac as well as dc components. Therefore, whenever even-order terms are present, their rectifying property will introduce constant currents even though the control voltage may be purely small-signal ac.

For this and subsequent discussions, small-signal is defined as being sufficiently small such that the nonlinear characteristic is adequately represented by a truncated Taylor series of no more than three terms. This definition is purposely vague. Whether a signal can be considered small-signal depends on the severity of the nonlinearity as well as the required accuracy both of which need to be determined on a case by case basis. Note also that the term small-signal refers to both ac and dc voltages and currents.

The polynomial for the small-signal $i(t)$, given by (2.6) has the same form as the original global one given by (2.2), however, the new coefficients do indeed represent the incremental conductances.

$$i(t) = g_1 V^1 + g_2 V^2 + g_3 V^3 + \cdots + g_k V^k \qquad (2.7)$$

$$where \; g_n = \frac{1}{n!} \frac{d^n I}{d V^n}\bigg|_{V=V_0} \qquad (2.8)$$

Note K is not necessarily equal to N. K, like N, is determined based on the maximum allowable or desired error between the raw and approximate data.

This need to perform a Taylor series expansion on the approximating function illustrates two points mentioned earlier:

1. The approximating function should have continuous derivatives. A Taylor series expansion doesn't exist without them.

2. The curve fitting technique should represent the data in a smooth manner. The incremental coefficients of the Taylor series will not represent the true nature of the nonlinearity about an operating point if the instantaneous slope of the approximating function varies radically.

3.0 POWER SERIES ANALYSIS

3.1 INTRODUCTION

Power series analysis is a relatively simple technique for analyzing weakly nonlinear, zero-memory systems. As the name implies, the nonlinear system is represented by a power series of the form:

$$y(t) = a_1 x + a_2 x^2 + a_3 x^3 + \cdots \tag{3.1}$$

$$y(t) = \sum_{n=1}^{\infty} a_n x^n \tag{3.2}$$

A system is said to have memory if the output depends on the past. Systems with energy storage devices (i.e. inductors, capacitors, spring moduli) have memory. The weakly nonlinear restriction comes from the need to represent the nonlinearity with a convergent power series. Additionally, if the nonlinearity cannot be adequately described by a limited number of terms (i.e. $N \leq 5$), the analysis becomes quite laborious. The restriction on memoryless systems stems from the fact that no provision is made to take into account events occurring in the past. It will be shown that when the necessary provisions are made the result is a Volterra series. It will also be shown that a power series is nothing more than a special case of the more general Volterra series. Although the restriction to memoryless systems does seriously limit its applicability, the power series approach is useful in some cases and does provide intuition into nonlinear systems.

3.2 GENERAL APPROACH

The power series method [8], [11], [25], [28] assumes that a system can be represented in terms of an isolated memoryless nonlinearity preceded and/or followed by linear network(s) as shown in Figure 3.1.

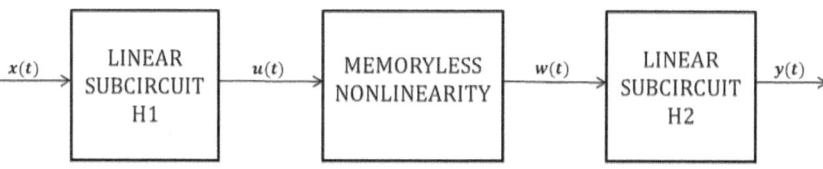

Figure 3.1 Power Series Analysis Model

In Figure 3.1 H_1 and H_2 represent all the linear frequency-dependent elements of the system. The nonlinearity between the two linear systems must have no frequency dependence and be represented by a power series. Note that no restrictions have been placed on H_1 and H_2 other than that they must be linear. They may represent filters or matching networks and as such will contain energy storage elements. Therefore, the system as a whole does contain memory. This statement does not contradict the requirement stated earlier of a memoryless nonlinearity as long as H_1 and H_2 are

27

completely independent of the nonlinear element. In other words, H_1 and H_2 cannot load the nonlinear element when all three are cascaded. Thus each block in Figure 3.1 can be treated as an individual isolated system whose input/output relationship can be determined independently of what drives or follows it. This requirement, difficult to meet in real life, illustrates the limitations of this approach.

Thus far no assumptions have been made which preclude the use of either time- or frequency-domain techniques to solve the given nonlinear system. Since H_1 and H_2 are linear networks, there exists an impulse response for each, $h_1(t)$ and $h_2(t)$, that completely defines their input/output relationship. Therefore, $u(t)$ can be found by convolving $h_1(t)$ with $x(t)$. The result is then substituted into the power series to obtain $w(t)$. Finally, $w(t)$ is convolved with $h_2(t)$ to achieve the final result $y(t)$. This method is obviously computationally intensive and provides little insight into the overall effect of the nonlinearity.

The usual motivation for analyzing a nonlinear system is to gain frequency-domain information (e.g. harmonic content, intermodulation distortion). Therefore, it is much more efficient to use frequency-domain techniques from the outset. An additional advantage to remaining in the frequency domain is that the transfer functions H_1 and H_2 can be determined by applying conventional linear analysis techniques in the s-plane instead of solving differential equations or performing inverse Laplace transforms.

3.3 SINUSOIDAL RESPONSE OF MEMORYLESS NONLINEAR SYSTEMS

The sinusoidal steady-state response of a nonlinear system [8] is usually of primary interest since most periodic signals occurring in communications can be decomposed into a sum of harmonically related sinusoids via a Fourier series. Another advantage in using sinusoids to characterize a nonlinear system is that it will be easy to determine what affect the nonlinearities have on the individual signal's amplitude and frequency.

It is generally easier to work with the exponential representation of a sinusoid as opposed to the trigonometric form. Therefore, the Euler form will be used throughout this thesis to represent sinusoidal signals:

$$x(t) = A\cos(w_0 t) = \sum_{q=-1}^{1} \frac{A}{2} e^{jqw_0 t} \, ; q \neq 0 \qquad (3.3)$$

Note the $q \neq 0$ will be assumed from here on and therefore, will not be explicitly stated.

As mentioned earlier a memoryless nonlinear system can be represented by a power series of the form:

$$y(t) = \sum_{n=1}^{\infty} Y_n(t) = \sum_{n=1}^{\infty} a_n x^n \qquad (3.4)$$

The first-order response ($n = 1$) to a sinusoid is trivial, namely:

$$Y_1(t) = a_1 \sum_{q=-1}^{1} \frac{A}{2} e^{jqw_0 t} \qquad (3.5)$$

The second-order response ($n = 2$) is a little more complicated:

$$Y_2(t) = a_2 [\sum_{q=-1}^{1} \frac{A}{2} e^{jqw_0 t}]^2 \qquad (3.6)$$

$$Y_2(t) = a_2 \sum_{q_1=-1}^{1} \frac{A}{2} e^{jq_1 w_0 t} \cdot \sum_{q_2=-1}^{1} \frac{A}{2} e^{jq_2 w_0 t}$$

$$Y_2(t) = \frac{A^2}{2^2} \sum_{q_1=-1}^{1} \sum_{q_2=-1}^{1} a_2 e^{j(q_1 w_0 + q_2 w_0)t} \qquad (3.7)$$

And finally, the nth-order response:

$$Y_n(t) = \frac{A^n}{2^n} \sum_{q_1=-1}^{1} \sum_{q_2=-1}^{1} \cdots \sum_{q_n=-1}^{1} a_n e^{j(q_1+q_2+\cdots+q_n)w_0 t} \qquad (3.8)$$

Therefore, the complete response $y(t)$ is given by:

$$y(t) = \sum_{n=1}^{\infty} Y_n(t) \tag{3.4}$$

$$y(t) = \sum_{n=1}^{\infty} \frac{A^n}{2^n} \sum_{q_1=-1}^{1} \sum_{q_2=-1}^{1} \cdots \sum_{q_n=-1}^{1} a_n e^{j(q_1+q_2+\cdots+q_n)w_0 t} \tag{3.9}$$

Equation 3.9 illustrates a few interesting points worth comment:

1. The most obvious feature of the response is that it contains frequency components not present at the input. This is a fundamental difference between linear and nonlinear systems.

2. The highest harmonic is equal to the highest order of the nonlinearity (i.e. the highest harmonic in $y_3(t)$ is $3w_0$).

3. Odd-order nonlinearities produce only odd-order harmonics and even-order nonlinearities produce only even-order harmonics.

4. The harmonic content is exacerbated at high input levels. Studying (3.9) shows that the power in each harmonic is proportional to the input power raised to the order of the harmonic. Therefore, for example, doubling the input power results in an 8-fold increase in the third harmonic power.

As mentioned earlier, limiting the nonlinearities to non-storage elements is unrealistic. In modern solid-state devices many of the nonlinearities result from the junction capacitances' dependences on node voltages. Consequently, power series analysis is, in general, impractical for such circuits. Nevertheless, some qualitative insight can be derived from its use.

3.4 LIMITATIONS

To illustrate the limitations imposed by power series analysis, the necessary approximations required to analyze a high-frequency MEtal Semiconductor Field Effect Transistor (MESFET) device will be demonstrated.

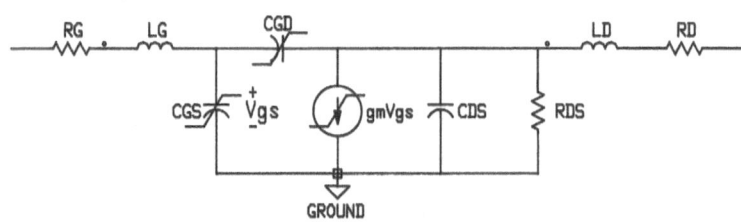

Figure 3.2 Complete MESFET Model

Of the three major sources of nonlinearities, the two involving the reversed-biased Schottky diodes modeled by C_{GS} and C_{GD} cannot, under any circumstances, be incorporated in a power series analysis. Additionally, in order to decouple the lone remaining nonlinearity, g_m, from the rest of the circuit, C_{GD} must be removed. As long as C_{GD} is connected it is impossible to manipulate the circuit in Figure 3.2 into the required form as shown in Figure 3.1. The loss of C_{GD} not only diminishes the model's ability to predict nonlinear behavior, it also significantly reduces its overall ability to model the high-frequency nature of the device's linear and nonlinear performance.

Therefore, to permit a power series analysis of the circuit in Figure 3.2, the circuit must be reduced into the following form (where C_{GS} is assumed linear):

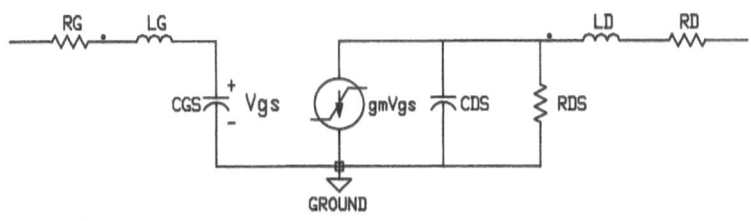

Figure 3.3 Modified MESFET Model

The circuit in Figure 3.3 is an idealized, single nonlinearity, low-frequency version of the original circuit. One way to extend the frequency range of the circuit in Figure 3.3 is to apply the Miller Theorem to C_{GD}. Under normal operating

33

conditions $(V_{GD} \geq 3)$ C_{GD} can be approximated as linear. Therefore, C_{GD} can be removed and replaced by the appropriate input and output Miller capacitances. Because the Miller capacitances use the midband linear gain, the further the actual gain deviates from its linearized value the larger the error introduced by this method. For some applications, however, even this procedure does not provide the necessary quantitative or qualitative information. For such cases, the Volterra series should be applied. This circuit will be reexamined later using Volterra series analysis.

4.0 VOLTERRA SERIES ANALYSIS

As mentioned earlier a Volterra series is a series of functionals. A functional is an operation which assigns to every function in a given set, or domain of functions, a unique real number [29].

For a time-invariant system it can be shown that the output of a nonlinear system with memory is related to the input by a functional series of the form:

$$y(t) = \sum_{n=1}^{\infty} \int_{-\infty}^{\infty} \cdots \int_{-\infty}^{\infty} \frac{h_n(\tau_1, \ldots, \tau_n)x(t-\tau_1) \cdots}{x(t-\tau_n)d\tau_1 \ldots d\tau_n} \qquad (4.1)$$

Another way of representing (4.1) is as follows:

$$y(t) = \sum_{n=1}^{\infty} Y_n[x(t)] \qquad (4.2)$$

where:

$$Y_n[x(t)] = \int_{-\infty}^{\infty} \cdots \int_{-\infty}^{\infty} \frac{h_n(\tau_1, \ldots, \tau_n)x(t-\tau_1) \cdots}{x(t-\tau_n)d\tau_1 \ldots d\tau_n} \qquad (4.3)$$

$Y_n[\cdot]$ is called the nth-order Volterra operator.

This functional series is a result of a generalized Taylor series expansion of a function first performed by Vito Volterra in the 1880's [22]. This generalization involves incorporating memory into power series expansions of nonlinear systems, thereby facilitating analysis of nonlinear systems with energy storage devices (i.e. capacitors and inductors). Therefore, a Volterra series can be viewed simply as a power series with memory [22]. More accurately stated, a power series is a Volterra series without memory. This latter statement can easily be demonstrated by showing how a power series can be rewritten in a Volterra-like notation [26].

$$y(t) = \sum_{n=1}^{\infty} a_n x^n \tag{4.4}$$

$$y(t) = a_1 x(t) + a_2 x^2(t) + a_3 x^3(t) + \cdots$$

$$
\begin{aligned}
y(t) = &\int_{-\infty}^{\infty} a_1 \delta(\tau_1) x(t - \tau_1) d\tau_1 \\
&+ \iint_{-\infty}^{\infty} a_2 \delta(\tau_1)\delta(\tau_2) x(t - \tau_1) x(t - \tau_2) d\tau_1 d\tau_2 \\
&+ \iiint_{-\infty}^{\infty} \begin{array}{l} a_3 \delta(\tau_1)\delta(\tau_2)\delta(\tau_3) x(t - \tau_1) x(t - \tau_2) \\ x(t - \tau_3) d\tau_1 d\tau_2 d\tau_3 + \cdots \end{array}
\end{aligned}
\tag{4.5}
$$

$$
\begin{aligned}
y(t) = &\int_{-\infty}^{\infty} h_1(\tau_1) x(t - \tau_1) d\tau_1 \\
&+ \iint_{-\infty}^{\infty} h_2(\tau_1,\tau_2) x(t - \tau_1) x(t - \tau_2) d\tau_1 d\tau_2 \\
&+ \iiint_{-\infty}^{\infty} \begin{array}{l} h_3(\tau_1,\tau_2,\tau_3) x(t - \tau_1) x(t - \tau_2) \\ x(t - \tau_3) d\tau_1 d\tau_2 d\tau_3 + \cdots \end{array}
\end{aligned}
\tag{4.6}
$$

where: $\quad h_1(\tau_1) = a_1 \delta(\tau_1)$
$\qquad\quad h_2(\tau_1, \tau_2) = a_2 \delta(\tau_1)\delta(\tau_2)$
$\qquad\quad h_3(\tau_1, \tau_2, \tau_3) = a_3 \delta(\tau_1)\delta(\tau_2)\delta(\tau_3)$

Notice that the system impulse responses are themselves impulses. This characteristic is a mathematical representation of a system without memory.

The power series nature of a Volterra series is easily demonstrated by multiplying the input by a constant 'a' so that the input becomes $ax(t)$ [23]. Substituting the new input into (4.2) results in a new output:

$$y(t) = \sum_{n=1}^{\infty} Y_n[ax(t)] \tag{4.7}$$

$$y(t) = \sum_{n=1}^{\infty} a^n Y_n[x(t)] \tag{4.8}$$

This new output is a power series in the constant 'a'. The memory handling capability is substantiated by the fact that the Volterra operators are convolution integrals and as such account for past and future (if noncausal) events when determining the response for a given time, t. The fact that the Volterra series is a generalization of a Taylor series results in the Volterra series suffering from the same limitations, namely it is not always convergent. As with a standard Taylor series, a Volterra series has a region of convergence associated with it. Therefore, when using a Volterra series there can exist inputs for which the series diverges and thus does not exist.

4.1 CONVERGENCE

Recall that the radius of convergence for a power series as represented by (4.4) is given by the d'Alambert ratio criterion [1]:

$$\frac{1}{R} = lim_{n \to \infty} \left| \frac{a_{n+1}}{a_n} \right|$$

This means that whenever $|x| < R$ the power series given by (4.4) converges absolutely. In an analogous manner a radius of convergence for a Volterra series can be defined [18].

Beginning with the standard form of the Volterra series representation a radius of convergence can be developed as follows:

$$y(t) = \sum_{n=1}^{\infty} \int_{-\infty}^{\infty} \dots \int_{-\infty}^{\infty} \frac{h_n(\tau_1, \dots, \tau_n)x(t - \tau_1) \dots}{x(t - \tau_n)d\tau_1 \dots d\tau_n} \qquad (4.1)$$

For the system to be stable the output must be bounded for a bounded input. With this in mind (4.1) can be rewritten:

Bounded input $=> |x(t)| < X$ where X is a finite number.

$$|y(t)| = \left| \sum_{n=1}^{\infty} Y_n(t) \right|$$

$$|y(t)| \le \sum_{n=1}^{\infty} \int_{-\infty}^{\infty} \cdots \int_{-\infty}^{\infty} \frac{|h_n(\tau_1, \ldots, \tau_n)||x(t-\tau_1)| \ldots |x(t-\tau_n)|}{d\tau_1 \ldots d\tau_n}$$

$$|y(t)| < \sum_{n=1}^{\infty} X^n \int_{-\infty}^{\infty} \cdots \int_{-\infty}^{\infty} |h_n(\tau_1, \ldots \tau_n)| d\tau_1 \ldots d\tau_n$$

$$|y(t)| < \sum_{n=1}^{\infty} g_n X^n$$

where $g_n = \int_{-\infty}^{\infty} \cdots \int_{-\infty}^{\infty} |h_n(\tau_1, \ldots \tau_n)| d\tau_1 \ldots d\tau_n$

Therefore, applying the d'Alambert criterion to the above equation yields the radius of convergence for a general Volterra series:

$$\frac{1}{R} = \lim_{n \to \infty} \left| \frac{g_{n+1}}{g_n} \right|$$

4.2 DEVELOPMENT OF VOLTERRA SERIES

In order to justify the earlier statements referring to a Volterra series as a generalized Taylor series as well as provide a more intuitive understanding of what a Volterra series actually is, a Volterra series for a general nonlinear system will be developed from a Taylor series. Although somewhat different in treatment, references [8], [26] and [28] did provide insight, guidance and in general laid the foundation for this section.

Note that this derivation is not intended to be mathematically rigorous but rather conceptual in nature. For more of a fundamental mathematical treatise on the subject refer to reference [5].

In order to develop a general system input/output relationship, a suitable basis function should be chosen. As in the case of linear systems, the input function will be represented by a superposition of impulse functions. Unlike the linear case, however, the response of a nonlinear system to a single impulse is not sufficient for complete system characterization. For a linear system, once the impulse response has been determined, the superposition principle can be used to construct the system response to any arbitrary input. The only requirement is that the input waveform can be represented by a series of basis functions that in the limit become impulse functions.

For a nonlinear system the superposition principle is no longer valid. As a consequence, the response of a nonlinear system to an impulse may be a nonlinear function of both its amplitude and time of occurrence. Therefore, if the input, $x(t)$, is decomposed into N basis functions, the output, $y(t)$, can be represented as a nonlinear function of N input variables.

The basis function used will be the rectangular pulse shown in Figure 4.1.

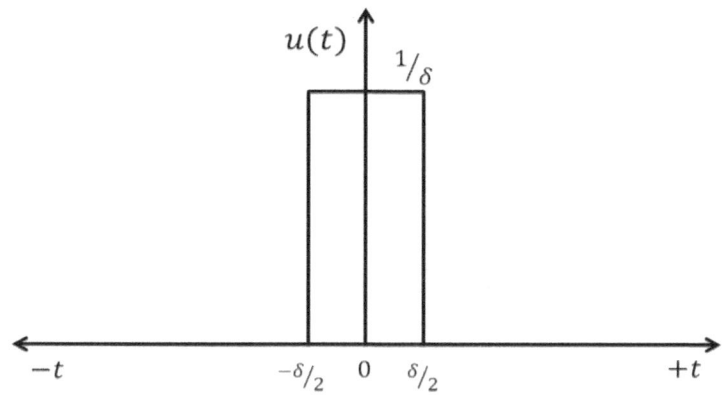

Figure 4.1 The basis function $U_\delta(t)$

Any waveform of interest can be approximated by a superposition of this function.

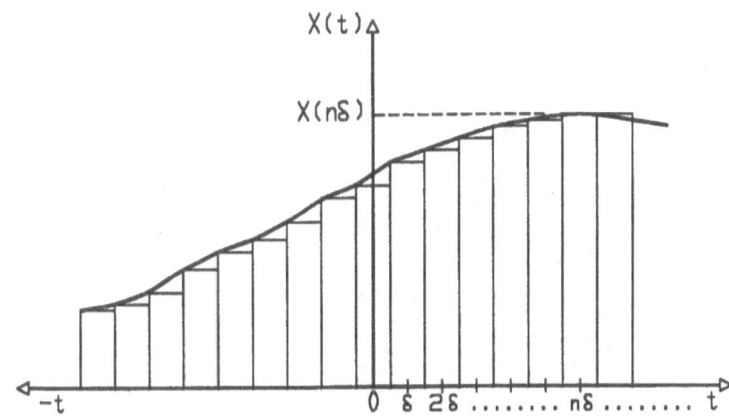

Figure 4.2 Input function $x(t)$ & approximation $x_\delta(t)$

In terms of the basis function $U_\delta(t)$, $x(t)$ can be expressed as:

$$x(t) = \lim_{\delta \to 0} x_\delta(t)$$

where $x_\delta(t) = \sum_{n=-N}^{N} \delta x(n\delta) U_\delta(t - n\delta)$ (4.9)

The effect of a single pulse will depend on its magnitude, $x(k\delta)$, its width, δ, its time of occurrence, $k\delta$, and the time, t. If time invariance is assumed, then only the time difference, $t - k\delta$, is relevant.

Consequently, the system's output at time t for a pulse occurring at $k\delta$ may be represented as follows:

$$y_{k\delta}(t) = f[x_{k\delta}(t)]$$ (4.10)

where $x_{k\delta}(t) = \delta x(k\delta) U_\delta(t - k\delta)$

When there is more than one input pulse, the response at time t will be a nonlinear function of each pulse's amplitude, width and time of occurrence.

The response for $2N + 1$ pulses is:

$$y_\delta(t) = f[x_{-N\delta}, \dots, x_{0\delta}, \dots, x_{N\delta}]$$

$$y_\delta(t) = f[X_\delta]; \text{ where } X_\delta = (x_{-N\delta}, \dots, x_{0\delta}, \dots, x_{N\delta})$$

Now expand $y_\delta(t)$ into an N-dimensional Taylor series about the origin as a function of the amplitudes; $x(-N\delta), \dots, x(0\delta), \dots, x(N\delta)$. Grouping terms having identical order yields $y_\delta(t)$ in the following form:

$$y_\delta(t) = \sum_{n_1=-N}^{N} \left. \frac{df(X_\delta)}{dx_{n_1\delta}} \right|_{X_\delta=0} \cdot x_{n_1\delta}$$

$$+ \sum_{n_1=-N}^{N} \sum_{n_2=-N}^{N} \left. \frac{d^2 f(X_\delta)}{dx_{n_1\delta} dx_{n_2\delta}} \right|_{X_\delta=0} \cdot x_{n_1\delta} x_{n_2\delta}$$

$$+ \sum_{n_1=-N}^{N} \sum_{n_2=-N}^{N} \sum_{n_3=-N}^{N} \left. \frac{d^3 f(X_\delta)}{dx_{n_1\delta} dx_{n_2\delta} dx_{n_3\delta}} \right|_{X_\delta=0} \cdot x_{n_1\delta} x_{n_2\delta} x_{n_3\delta}$$

$$+ \dots \tag{4.12}$$

Substituting the expressions for $x_{n_1\delta}, x_{n_2\delta}, x_{n_3\delta} \dots$ into (4.12) can be rewritten as follows:

$$y_\delta(t) = \sum_{n_1=-N}^{N} \left. \frac{df(X_\delta)}{dx_{n_1\delta}} \right|_{X_\delta=0} \cdot x(n_1\delta) U_\delta(t - n_1\delta)\delta$$

$$+ \sum_{n_1=-N}^{N} \sum_{n_2=-N}^{N} \left. \frac{d^2 f(X_\delta)}{dx_{n_1\delta} dx_{n_2\delta}} \right|_{X_\delta=0} \cdot x(n_1\delta) U_\delta(t - n_1\delta)\delta$$

$$\cdot x(n_2\delta) U_\delta(t - n_2\delta)\delta$$

$$+ \sum_{n_1=-N}^{N} \sum_{n_2=-N}^{N} \sum_{n_3=-N}^{N} \left. \frac{d^3 f(X_\delta)}{dx_{n_1\delta} dx_{n_2\delta} dx_{n_3\delta}} \right|_{X_\delta=0} \cdot x(n_1\delta)$$

$$\cdot U_\delta(t - n_1\delta)\delta \cdot x(n_2\delta) U_\delta(t - n_2\delta)\delta \cdot x(n_3\delta) U_\delta(t - n_3\delta)\delta$$

$$+ \dots$$

$$\tag{4.13}$$

Equation 4.13 can be further simplified by introducing a new function $G_n(\cdot)$; where $G_n(\cdot)$ is the nth Taylor coefficient multiplied by the basis functions:

$$G_1(t - n_1\delta) = \frac{df(X_\delta)}{dx_{n_1\delta}}\bigg|_{X_\delta=0} \cdot U_\delta(t - n_1\delta) \tag{4.14}$$

$$G_2(t - n_1\delta, t - n_2\delta) =$$
$$\frac{d^2f(X_\delta)}{dx_{n_1\delta}dx_{n_2\delta}}\bigg|_{X_\delta=0} \cdot U_\delta(t - n_1\delta)U_\delta(t - n_2\delta) \tag{4.15}$$

$$G_3(t - n_1\delta, t - n_2\delta, t - n_3\delta) =$$
$$\frac{d^3f(X_\delta)}{dx_{n_1\delta}dx_{n_2\delta}dx_{n_3\delta}}\bigg|_{X_\delta=0} \cdot U_\delta(t - n_1\delta)U_\delta(t - n_2\delta)U_\delta(t - n_3\delta)$$
$$\tag{4.16}$$

Therefore, (4.13) becomes:

$$y_\delta(t) = \sum_{n_1=-N}^{N} G_1(t - n_1\delta)x(n_1\delta)\delta$$
$$+ \sum_{n_1=-N}^{N} \sum_{n_2=-N}^{N} G_2(t - n_1\delta, t - n_2\delta)x(n_1\delta)\delta \cdot x(n_2\delta)\delta$$
$$+ \sum_{n_1=-N}^{N} \sum_{n_2=-N}^{N} \sum_{n_3=-N}^{N} G_3(t - n_1\delta, t - n_2\delta, t - n_3\delta)$$
$$\cdot x(n_1\delta)\delta \cdot x(n_2\delta)\delta \cdot x(n_3\delta)\delta + \cdots \tag{4.17}$$

At this point the output response, $y(t)$ can be determined from the approximation function $y_\delta(t)$ by letting $\delta \to 0$.

$$y(t) = \lim_{\delta \to 0} y_\delta(t)$$

Note that as $\delta \to 0$ the width of the pulses becomes infinitesimal so that $\delta \to d\sigma, N \to \infty, k\delta \to \sigma$ and $G \to h$. The result of this operation is that the summations become integrals and the Taylor coefficients multiplied by the basis function become impulse responses.

$$y(t) = \int_{-\infty}^{\infty} h_1(t - \sigma_1)x(\sigma_1)d\sigma_1$$
$$+ \iint_{-\infty}^{\infty} h_2(t - \sigma_1, t - \sigma_2)x(\sigma_1)x(\sigma_2)d\sigma_1 d\sigma_2$$
$$+ \iiint_{-\infty}^{\infty} h_3(t - \sigma_1, t - \sigma_2, t - \sigma_3) x(\sigma_1)x(\sigma_2)x(\sigma_3)d\sigma_1 d\sigma_2 d\sigma_3$$

$$(4.18)$$

where $h_n(\cdot) = \lim_{\delta \to \infty} G_n(\cdot)$

By performing a simple change of variable operation: $\tau_n = t - \sigma_n$ for $n = 1,2, \dots, \infty$, (4.18) can be transformed into the standard form of a Volterra series as described earlier in this chapter:

$$y(t) = \int_{-\infty}^{\infty} h_1(\tau_1)x(t - \tau_1)d\tau_1$$
$$+ \iint_{-\infty}^{\infty} h_2(\tau_1, \tau_2)x(t - \tau_1)x(t - \tau_2)d\tau_1 d\tau_2$$
$$+ \iiint_{-\infty}^{\infty} h_3(\tau_1, \tau_2, \tau_3)x(t - \tau_1)x(t - \tau_2) x(t - \tau_3)d\tau_1 d\tau_2 d\tau_3$$

$$(4.6)$$

Equation 4.6 can be rewritten in terms of Volterra operators:

$$y(t) = Y_1[x(t)] + Y_2[x(t)] + Y_3[x(t)] \dots \qquad (4.19)$$

At this point it is useful to illustrate a few points regarding both (4.6) and (4.19):

1. $Y_1[x(t)]$, the first-order Volterra operator, is nothing more than the standard convolution integral for linear time-invariant systems. This follows from the fact that linear and nonlinear system theory should converge for a linear network.

2. By analogy to Y_1; Y_2, Y_3 and in general Y_N represent two-, three- and n-dimensional convolution integrals respectively. The functions h_1, h_2, h_3 and h_n are called the first, second, third and nth-order Volterra kernels. Note that h_1 is the impulse response for a linear time-invariant system. Again by analogy, h_2, h_3 and h_n can be interpreted as second-, third-, and nth-order impulse responses.

3. A Volterra series is a series of multilinear operators. An operator is multilinear if it is linear in each argument, or dimension, when all others are held fixed. This statement can easily be illustrated by examining the nth-order Volterra operator:

$$Y_n[x(t)] = \int_{-\infty}^{\infty} \cdots \int_{-\infty}^{\infty} \frac{h_n(\tau_1, \tau_2, \ldots, \tau_n) x(t - \tau_1) x(t - \tau_2) \cdots}{x(t - \tau_n) d\tau_1 d\tau_2 \ldots d\tau_n}$$

(4.3)

If all argument but one, $x_i(t - \tau_i)$, are held fixed, the multidimensional convolution integral collapses into (4.20) where K is a constant multiplying a one-dimensional convolution integral. The constant, K, results from the $n - 1$ arguments being held constant and therefore, pulled out of the integral.

$$Y_i[x(t)] = K \int_{-\infty}^{\infty} h(\tau_i)x(t - \tau_i)d\tau_i \qquad (4.20)$$

Equation 4.20 has the form of a standard linear time-invariant convolution integral and is linear by definition. Note that $h(\tau_i)$ is not necessarily $h_1(\tau_i)$ but it is first-order.

Another way of demonstrating the multilinear nature of Volterra operators is to remember that they were derived from a multidimensional Taylor series expansion, which is itself multilinear.

This last statement is extremely important because it defines the type of nonlinear system that can be represented by a Volterra series. Any nonlinear system whose nonlinearities are expandable in a convergent multilinear series can be represented by a Volterra series and is therefore, referred to as a Volterra system. Note that when the nonlinearity is a

47

function of only one variable the multilinear series collapses into a power series.

4.3 FIRST-ORDER VOLTERRA SYSTEMS

In this section the various properties of Volterra operators are examined. The theory for first-order Volterra systems, in other words linear systems, is first reviewed. In doing so it will be shown that the theory for higher-order Volterra systems can be viewed as a generalization of the theory for a first-order Volterra system. In subsequent sections, these concepts will be extended to higher-order systems. By building on a foundation of familiar work, a qualitative as well as quantitative insight into nonlinear systems can be gained.

4.3.1 FIRST-ORDER IMPULSE RESPONSE

Recall that (4.1) represents the Taylor series expansion of a general nonlinear system. A first-order Volterra system is defined by the first term in (4.1):

$$y(t) = \int_{-\infty}^{\infty} h_1(\tau)x(t-\tau)d\tau \qquad (4.21)$$

Where $h_1(\tau)$ is referred to as the first-order impulse response, or simply the impulse response. For a first-order system (linear) the impulse response is all that is required to completely characterize the system. It will be shown that for higher-order systems the simple first-order impulse response is no longer adequate and that higher-order impulse responses are required.

4.3.2 CAUSALITY

A system is said to be causal if, for any input, the response at any instant of time does not depend upon future inputs. Naturally all real systems are causal. How these constraints affect the impulse response can be seen by examining (4.22). Note that the integrand in (4.22) is the same as in (4.21) with $\sigma = t - \tau$. Since t denotes the present time, whenever σ exceeds t the future is being addressed. Therefore, for a causal system the following integral must be equal to zero:

$$y(t) = \int_t^\infty x(\sigma)h_1(t - \sigma)d\sigma = 0 \qquad (4.22)$$

For this equation to hold for any function $x(\sigma)$, the kernel $h_1(\cdot)$ must be identically equal to zero whenever $\sigma > t$. Mathematically phrased:

$$h_1(t - \sigma) = 0; \quad for\ \sigma > t\ or\ t - \sigma < 0$$
$$or$$
$$h_1(\tau) = 0; \quad for\ \tau < 0 \qquad (4.23)$$

(4.23) is both a necessary and sufficient condition for a linear time invariant system to be causal.

4.3.3 STABILITY

Generally when discussing Volterra systems, a stable system is defined as one in which for every bounded input the output is also bounded. This is referred to as "bounded input, bounded output", or BIBO, stability.

Mathematically stated:

For $|x(t)| < M$ where M is some finite number $|y(t)| < \infty$

A system will be BIBO stable if

$$\int_{-\infty}^{\infty} |h_1(\tau)| \, d\tau < \infty \tag{4.24}$$

Equation 4.24 is easily proven as follows:

$$|y(t)| = \left| \int_{-\infty}^{\infty} h_1(\tau) x(t - \tau) d\tau \right|$$

$$|y(t)| \leq \int_{-\infty}^{\infty} |h_1(\tau)| |x(t - \tau)| d\tau$$

$$|y(t)| < M \int_{-\infty}^{\infty} |h_1(\tau)| \, d\tau$$

Therefore,

$$\int_{-\infty}^{\infty} |h_1(\tau)| \, d\tau < \infty \implies |y(t)| < \infty$$

Note this proves (4.24) is sufficient to guarantee BIBO. Equation (4.24) can also be shown to be necessary [23].

4.3.4 FIRST-ORDER VOLTERRA TRANSFER FUNCTION

Recall that the Fourier transform of the impulse response is referred to as the transfer function. The mathematical definition is given below:

$$H(\omega) = \int_{-\infty}^{\infty} h(\tau) e^{-j\omega\tau} \, d\tau \tag{4.25}$$

$$h(t) = \frac{1}{2\pi} \int_{-\infty}^{\infty} H(\omega) e^{j\omega t} \, d\omega \tag{4.26}$$

Note that being BIBO stable is a sufficient condition for the system's transfer function $H(\omega)$ to exist:

$$|H(\omega)| \leq \int_{-\infty}^{\infty} |h(\tau)| |e^{-j\omega\tau}| \, d\tau$$

$$|H(\omega)| \leq \int_{-\infty}^{\infty} |h(\tau)| \, d\tau$$

Therefore, if the system is stable, $|H(\omega)|$ will be finite for all ω.

4.3.5 SINUSOIDAL RESPONSE OF FIRST-ORDER VOLTERRA SYSTEMS

Many real periodic signals can be represented by a Fourier series of the form:

$$x(t) = \sum_{n=-\infty}^{\infty} c_n e^{jn\omega t} \tag{4.27}$$

The output of a linear time-invariant system can then be expressed by applying superposition:

$$y(t) = \sum_{n=-\infty}^{\infty} c_n H(n\omega) e^{jn\omega t} \tag{4.28}$$

where $H(\cdot)$ is the system transfer function.

Applying (4.27) for a sinusoidal input, $x(t) = A\cos(\omega_0 t)$ yields:

$$x(t) = A\cos(\omega_0 t) = \frac{A}{2} e^{j\omega_0 t} + \frac{A}{2} e^{-j\omega_0 t} \tag{4.29}$$

$$x(t) = \sum_{n=-1}^{1} \frac{A}{2} e^{jn\omega_0 t} \quad ; n \neq 0 \text{ and } c_n = \frac{A}{2} \tag{4.30}$$

Note that (4.30) is simply the Euler form of a sinusoidal signal as described in Chapter 3. Again $n \neq 0$ will be assumed. Using (4.28), the output response, $y(t)$, is determined:

$$y(t) = \sum_{n=-1}^{1} \frac{A}{2} H(n\omega_0) e^{jn\omega_0 t} \tag{4.31}$$

$$y(t) = \frac{A}{2} H(\omega_0) e^{j\omega_0 t} + \frac{A}{2} H(-\omega_0) e^{-j\omega_0 t} \tag{4.32}$$

$$y(t) = \frac{A}{2} H(\omega_0) e^{j\omega_0 t} + \frac{A}{2} H^*(\omega_0) e^{-j\omega_0 t} \tag{4.33}$$

Recall that since $h(t)$ is assumed to be a real function of time:

$$H(-\omega_0) = H^*(\omega_0)$$

Now let: $H(\omega_0) = |H(\omega_0)| e^{j\theta}$ and $H^*(\omega_0) = |H(\omega_0)| e^{-j\theta}$

Therefore, (4.32) becomes:

$$y(t) = \frac{A}{2} |H(\omega_0)| \left(e^{j(\omega_0 t + \theta)} + e^{-j(\omega_0 t + \theta)} \right) \tag{4.34}$$

$$y(t) = A|H(\omega_0)| \cos(\omega_0 t + \theta) \tag{4.35}$$

Based on (4.35) the following observations can be made:
1. The output contains no frequency components that were not also present at the input.
2. The output is shifted in phase by the system phase shift, θ and the amplitude is changed by the system gain, $|H(\omega_0)|$.
3. The system gain and phase shift completely characterize a stable, linear, time-invariant system.

4.4 SECOND-ORDER VOLTERRA SYSTEMS

4.4.1 SECOND-ORDER IMPULSE RESPONSE

When dealing with higher-order systems, the physical interpretation becomes somewhat more difficult. In the first-order case, the kernel $h(\tau)$, is the response of the system to an impulse and completely characterizes the system. In higher-order systems this is no longer true. This point is easily shown by finding the impulse response of a second-order Volterra system. The second-order Volterra operator for the system is:

$$Y_2[x(t)] = \iint_{-\infty}^{\infty} h_2(\tau_1, \tau_2) x(t - \tau_1) x(t - \tau_2) d\tau_1 d\tau_2 \qquad (4.36)$$

Therefore, the system impulse response is found by letting $x(t) = \delta(t)$.

$$Y_2[\delta(t)] = \iint_{-\infty}^{\infty} h_2(\tau_1, \tau_2) \delta(t - \tau_1) \delta(t - \tau_2) d\tau_1 d\tau_2 \qquad (4.37)$$

$$Y_2[\delta(t)] = h_2(t, t) \qquad (4.38)$$

The above result shows that the impulse response of a second-order system only characterizes the system along the vector $\tau_1 = \tau_2$ or in other words, along the diagonal of the second-order system response. To illustrate this last statement an arbitrary second-order system response is sketched in Figure 4.3.

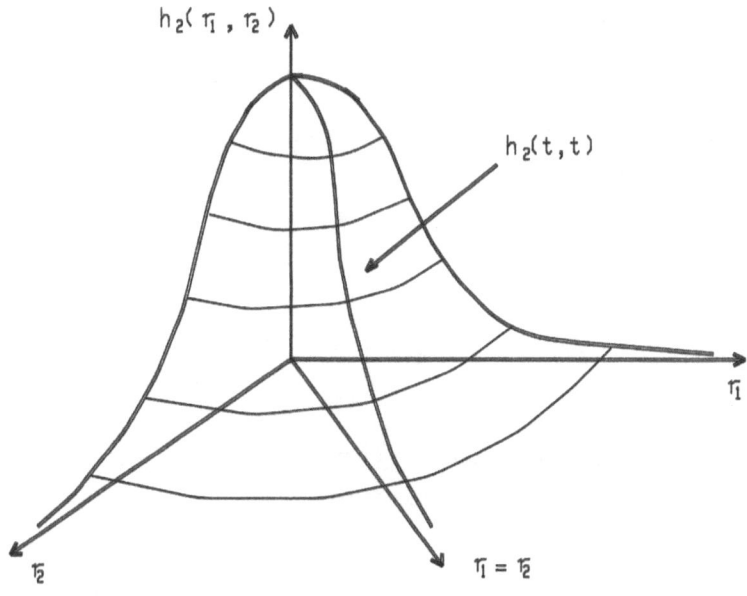

Figure 4.3 Second-order impulse response

Determination of the off-diagonal values of the system response requires a second-order impulse response. The second-order impulse response can be found from two times the bilinear term that results from the application of two impulses occurring at different times. The "two times" is a result of the assumption that the kernel, $h_2(\tau_1, \tau_2)$, is symmetric with respect to its arguments τ_1 and τ_2. It will be shown in the Section 4.4.2 that this assumption results in no loss of generality.

55

The procedure for determining the second-order impulse response is as follows [23]:

Let: $x(t) = \delta(t - T_1) + \delta(t - T_2)$ (4.39)

where T_1 and T_2 are arbitrary delays.

Therefore,

$$Y_2[x(t)] = Y_2[\delta(t - T_1) + \delta(t - T_2)]$$ (4.40)

$$Y_2(t) = \iint_{-\infty}^{\infty} \frac{h_2(\tau_1, \tau_2)[\delta(t - T_1 - \tau_1) + \delta(t - T_2 - \tau_1)] \cdot}{[\delta(t - T_1 - \tau_2) + \delta(t - T_2 - \tau_2)]\delta\tau_1 d\tau_2}$$

(4.41)

$$Y_2(t) = \iint_{-\infty}^{\infty} h_2(\tau_1, \tau_2) \, \delta(t - T_1 - \tau_1)\delta(t - T_1 - \tau_2)\delta\tau_1 d\tau_2$$

$$+ \iint_{-\infty}^{\infty} h_2(\tau_1, \tau_2) \, \delta(t - T_2 - \tau_1)\delta(t - T_2 - \tau_2)\delta\tau_1 d\tau_2$$

$$+2 \iint_{-\infty}^{\infty} h_2(\tau_1, \tau_2) \, \delta(t - T_1 - \tau_1)\delta(t - T_2 - \tau_2)\delta\tau_1 d\tau_2$$

(4.42)

$$Y_2(t) = h_2(t - T_1, t - T_1) + h_2(t - T_2, t - T_2)$$
$$+2h_2(t - T_1, t - T_2)$$ (4.43)

Note that the first two terms in (4.43) again only describe the system impulse response along the $\tau_1 = \tau_2$ vector; whereas the last term, the bilinear term, describes the system impulse response anywhere in the $\tau_1 - \tau_2$ plane. By varying T_1 and T_2 in the bilinear expression, the system impulse response can be completely characterized. Therefore, the second-order impulse response for symmetric kernels can be expressed as follows:

$$g_2(t) = h_2(t - T_1, t - T_2) \tag{4.44}$$

In other words, (4.44) completely specifies the second-order impulse response. Note that (4.44) is the bilinear term divided by two. The division by two is necessary to remove the two generated by assuming symmetric kernels. The other terms in (4.43) are merely subsets of the more general bilinear term and as such are superfluous when determining the second-order impulse response. They are, however, necessary when determining the systems response to two impulses. This derivation shows that there is a subtle yet critical distinction between the impulse response of a higher-order Volterra system and the higher-order impulse response of a Volterra system. The latter is the higher-order analog to the impulse response of a linear system.

4.4.2 KERNEL SYMMETRIZATION

A second-order kernel is symmetric if $h_2(\tau_1, \tau_2) = h_2(\tau_2, \tau_1)$. While Volterra kernels are not necessarily symmetric, there are two main reasons to work with symmetric kernels. First, the use of symmetric kernels greatly simplifies analysis since the order of the τ's becomes unimportant. This results in expressions that are easier to handle and reduces the amount of calculation. The second reason is that a symmetric kernel can be shown to be unique whereas asymmetric kernels are not. The advantage of determining unique and hence symmetric kernels was seen in the previous section where by assuming symmetric kernels, only one bilinear term needed to be determined. Fortunately, there exists a procedure [23] that symmetrizes any asymmetric kernel.

The procedure begins with (4.36) the general second-order operator.

$$Y_2[x(t)] = \iint_{-\infty}^{\infty} h_2(\tau_1, \tau_2)x(t - \tau_1)x(t - \tau_2)d\tau_1 d\tau_2 \qquad (4.36)$$

For this example $h_2(\tau_1, \tau_2)$ is assumed to be asymmetric. Since τ_1, and τ_2, are simply dummy integration variables and the integration is over the entire argument range of the kernel, the result, $Y_2[x(t)]$, is the same regardless of their order. Therefore, although $h_2(\tau_1, \tau_2) \neq h_2(\tau_2, \tau_1)$ the result, $Y_2[x(t)]$, is unaffected by the order of τ_1, and τ_2. Thus, the second-order system is fully characterized by either $h_2(\tau_1, \tau_2)$ or $h_2(\tau_2, \tau_1)$ and because of this property can be made

symmetric by simply adding the two results and dividing by two.

$$Y_2[x(t)] = \frac{1}{2}\left[\begin{array}{l} \iint_{-\infty}^{\infty} h_2(\tau_1, \tau_2)x(t - \tau_1)x(t - \tau_2)d\tau_1 d\tau_2 + \\ \iint_{-\infty}^{\infty} h_2(\tau_2, \tau_1)x(t - \tau_2)x(t - \tau_1)d\tau_2 d\tau_1 \end{array}\right]$$

(4.45)

$$Y_2[x(t)] = \frac{1}{2}\left[\begin{array}{c} \iint_{-\infty}^{\infty}[h_2(\tau_1, \tau_2) + h_2(\tau_2, \tau_1)]x(t - \tau_1) \\ x(t - \tau_2)d\tau_1 d\tau_2 \end{array}\right]$$

(4.46)

Therefore,

$$h_2^{sym}(\tau_1, \tau_2) = \frac{1}{2}[h_2(\tau_1, \tau_2) + h_2(\tau_2, \tau_1)]$$

(4.47)

An intuitive justification for symmetrizing kernels relies on the fact that for a second-order system an impulse pair is indistinguishable from the same impulse pair presented in reverse order.

This procedure applies to any asymmetric kernel, so there is no loss of generality by using exclusively symmetric kernels.

4.4.3 CAUSALITY

The same definition for causality used for linear systems applies to higher-order systems as well. The response of a causal system does not depend on the future. This imposes constraints on the second-order Volterra kernel that are determined below.

$$Y_2[x(t)] = \iint_{-\infty}^{\infty} h_2(\tau_1, \tau_2) x(t - \tau_1) x(t - \tau_2) d\tau_1 d\tau_2 \qquad (4.36)$$

Let: $\sigma_1 = t - \tau_1$ and $\sigma_2 = t - \tau_2$

$$Y_2[x(t)] = \iint_{-\infty}^{\infty} h_2(t - \sigma_1, t - \sigma_2) x(\sigma_1) x(\sigma_2) d\sigma_1 d\sigma_2 \qquad (4.48)$$

Because t denotes the present time, whenever σ_1 or σ_2 exceeds t the future is being depicted. Therefore, for causality to hold the following equality must hold for any input $x(t)$.

$$Y_2[x(t)] = \iint_{t}^{\infty} h_2(t - \sigma_1, t - \sigma_2) x(\sigma_1) x(\sigma_2) d\sigma_1 d\sigma_2 = 0$$

Therefore, $h_2(t - \sigma_1, t - \sigma_2) = 0$ for $\sigma_1 > t$ and/or $\sigma_2 > t$

This condition can be stated in a simpler form by substituting:

$\tau_1 = t - \sigma_1$ and $\tau_2 = t - \sigma_2$

The condition becomes:

$$h_2(\tau_1, \tau_2) = 0 \quad for\ \tau_1 < 0\ and/or\ \tau_2 < 0 \qquad (4.49)$$

In other words, $h_2(\tau_1, \tau_2)$ can only exist (i.e. be nonzero) in the space defined by the intersection of the planes $\tau_1 > 0$ and $\tau_2 > 0$. See Figure 4.3.

4.4.4 STABILITY

Again the "BIBO" criterion for stability is selected. Based on this criterion a sufficient condition for stability can be shown to be [23]:

$$\iint_{-\infty}^{\infty} |h_2(\tau_1, \tau_2)| \, d\tau_1 d\tau_2 < \infty \tag{4.50}$$

Equation 4.50 can be derived as follows:

BIBO $\Rightarrow |Y_2(t)| < \infty$ for $|x(t)| < M$; where M is some finite number.

$$\Rightarrow |Y_2[x(t)]| = \left| \iint_{-\infty}^{\infty} h_2(\tau_1, \tau_2) x(t - \tau_1) x(t - \tau_2) d\tau_1 d\tau_2 \right| < \infty$$

but:

$$|Y_2[x(t)]| \leq \iint_{-\infty}^{\infty} |h_2(\tau_1, \tau_2)| |x(t - \tau_1)| |x(t - \tau_2)| \, d\tau_1 d\tau_2 < \infty$$

or:

$$|Y_2[x(t)]| \leq M^2 \iint_{-\infty}^{\infty} |h_2(\tau_1, \tau_2)| d\tau_1 d\tau_2 < \infty$$

Note this proves the condition to be sufficient, not necessary.

Examples can be found for which the above condition is not necessary [22]. Recall that for linear time invariant systems the first-order analog to (4.50) was both necessary and sufficient.

4.4.5 SECOND-ORDER VOLTERRA TRANSFER FUNCTION

In a manner similar to the linear case, the second-order Volterra transfer function is simply the Fourier transform of the second-order Volterra kernel (impulse response). The only difference is that in the second-order case the impulse response is a function of two variables and therefore, requires a two-dimensional Fourier transform.

A two-dimensional Fourier transform can be viewed as a one-dimensional transform performed on one variable with the other held fixed, followed by another one-dimensional transform performed on the previously fixed variable [23].

The process is demonstrated below.

1. $h_2(\tau_1, \tau_2)$ is the function to be transformed. Hold τ_2 fixed and perform a one-dimensional Fourier transform on $h_2(\tau_1, \tau_2)$ with respect to τ_1.

$$H_1(\omega_1, \tau_2) = \int_{-\infty}^{\infty} h_2(\tau_1, \tau_2)e^{-j\omega_1\tau_1}d\tau_1 \qquad (4.51)$$

2. Now hold ω_1 fixed and perform a one-dimensional Fourier transform on $H_1(\omega_1, \tau_2)$ with respect to τ_2.

$$H_2(\omega_1, \omega_2) = \int_{-\infty}^{\infty} H_1(\omega_1, \tau_2) e^{-j\omega_2 \tau_2 \omega} d\tau_2 \qquad (4.52)$$

3. Now substitute (4.51) into (4.52) to yield the form of a two-dimensional Fourier transform.

$$H_2(\omega_1, \omega_2) = \iint_{-\infty}^{\infty} h_2(\tau_1, \tau_2) e^{-j(\omega_1 \tau_1 + \omega_2 \tau_2)} d\tau_1 d\tau_2 \qquad (4.53)$$

To find the inverse Fourier transform the same technique, two successive one-dimensional inverse Fourier transforms, can be applied in the reverse direction.

$$h_2(\tau_1, \tau_2) = \frac{1}{(2\pi)^2} \iint_{-\infty}^{\infty} H_2(\omega_1, \omega_2) e^{j(\omega_1 \tau_1 + \omega_2 \tau_2)} d\omega_1 d\omega_2 \qquad (4.54)$$

$H_2(\omega_1, \omega_2)$ is known as the second-order Volterra transfer function and characterizes the second-order system response in the frequency domain.

Note that $H_2(\omega_1, \omega_2)$ is symmetric with respect to its arguments if $h_2(\tau_1, \tau_2)$ is symmetric.

As in the linear case, the existence of the Fourier transform of the second-order Volterra kernel $h_2(\tau_1, \tau_2)$ is insured by the "BIBO" condition expressed in (4.50).

4.4.6 SINUSOIDAL RESPONSE OF SECOND-ORDER VOLTERRA SYSTEMS

The response of a system described by a second-order Volterra series to a sinusoid will be examined because of its importance in electrical engineering.

Let: $x(t) = A \cos(w_0 t) = \sum_{q=-1}^{1} \frac{A}{2} e^{jq w_0 t} \, ; q \neq 0$ \hfill (3.3)

Equation 4.28 is no longer valid because it depends on superposition. Therefore, the response, $Y_2(t)$, of a second-order system will be found by substituting (3.3) into (4.36).

$$Y_2[x(t)] = Y_2 \left[\sum_{q=-1}^{1} \frac{A}{2} e^{jq w_0 t} \right]$$

$$Y_2[x(t)] = \frac{A^2}{2^2} \iint_{-\infty}^{\infty} \frac{h_2(\tau_1 \tau_2) \left[\sum_{q_1=-1}^{1} e^{j q_1 w_0 (t-\tau_1)} \right]}{ \cdot \left[\sum_{q_2=-1}^{1} e^{j q_2 w_0 (t-\tau_2)} \right] d\tau_1 d\tau_2}$$

$$Y_2[x(t)] = \frac{A^2}{2^2} \sum_{q_1=-1}^{1} \sum_{q_2=-1}^{1} e^{j(q_1 w_0 + q_2 w_0)t} \iint_{-\infty}^{\infty} h_2(\tau_1, \tau_2)$$
$$\cdot e^{-j(q_1 w_0 \tau_1 + q_2 w_0 \tau_2)} \, d\tau_1 d\tau_2$$

\hfill (4.55)

Recall from the previous section that the integral in (4.55) is nothing more than the two-dimensional Fourier transform of $h_2(\tau_1, \tau_2)$.

64

Therefore, (4.55) can be rewritten more compactly as:

$$Y_2[x(t)] = \frac{A^2}{2^2} \sum_{q_1=-1}^{1} \sum_{q_2=-1}^{1} e^{j(q_1 w_0 + q_2 w_0)t} H_2(q_1 \omega_0, q_2 \omega_0)$$

(4.56)

Note the similarity of (4.56) to (3.7), the corresponding equation for a memoryless nonlinear system. Their similarities and differences will be discussed in detail in the next section.

Expanding (4.56) further yields:

$$Y_2[x(t)] = \frac{A^2}{2^2} e^{j2w_0 t} H_2(\omega_0, \omega_0) + \frac{A^2}{2^2} e^{-j2w_0 t} H_2(-\omega_0, -\omega_0)$$

$$+ \frac{A^2}{2^2} H_2(\omega_0, -\omega_0) + \frac{A^2}{2^2} H_2(-\omega_0, \omega_0)$$

(4.57)

Note: $H_2(\omega_0, \omega_0) = H_2^*(-\omega_0, -\omega_0)$

and,

$H_2(\omega_0, -\omega_0) = H_2^*(-\omega_0, \omega_0)$

Using these relationships, (4.57) can be expressed in the following form:

$$Y_2[x(t)] = \frac{A^2}{2} |H_2(\omega_0, \omega_0)| \frac{1}{2} \left[e^{j(2\omega_0 t + \theta_1)} + e^{-j(2\omega_0 t + \theta_1)} \right]$$

$$+ \frac{A^2}{2} |H_2(\omega_0, -\omega_0)| \frac{1}{2} \left[e^{j\theta_2} + e^{-j\theta_2} \right]$$

(4.58)

where:

$\theta_1 = angle[H_2(\omega_0, \omega_0)] \ and \ \theta_2 = angle[H_2(\omega_0, -\omega_0)]$

65

Finally:

$$Y_2[x(t)] = \frac{A^2}{2}|H_2(\omega_0, \omega_0)|\cos(2\omega_0 t + \theta_1)$$

$$+ \frac{A^2}{2}|H_2(\omega_0, -\omega_0)|\cos(\theta_2) \tag{4.59}$$

Equation 4.59 shows that the sinusoidal response of a second-order Volterra system contains a term at dc and one at twice the input sinusoidal frequency. This result stems from the definition of a nonlinear system namely; a nonlinear system doesn't obey superposition. Note that the output contains frequencies that are related to, but not equal to, the frequency of the input signal.

It follows that the second-order system response to a single sinusoid does not completely characterize the system as it does in the linear case. In order to gain a better understanding refer to Figure 4.4, a sketch of the absolute value of an arbitrary second-order transfer function as a function of ω_1 and ω_2. The amplitude of the $2\omega_0$ term in (4.59) is proportional to the height of the surface above the $\omega_1 = \omega_2$ line in the $\omega_1 - \omega_2$ plane. The constant term is proportional to the height of the surface above the $\omega_1 = -\omega_2$ line in the same plane. Therefore, (4.59), the sinusoidal response of a second-order Volterra system, only characterizes the system along 2 lines in the $w_1 - w_2$ plane. In order to characterize the entire system a method for determining the off-diagonal values is needed.

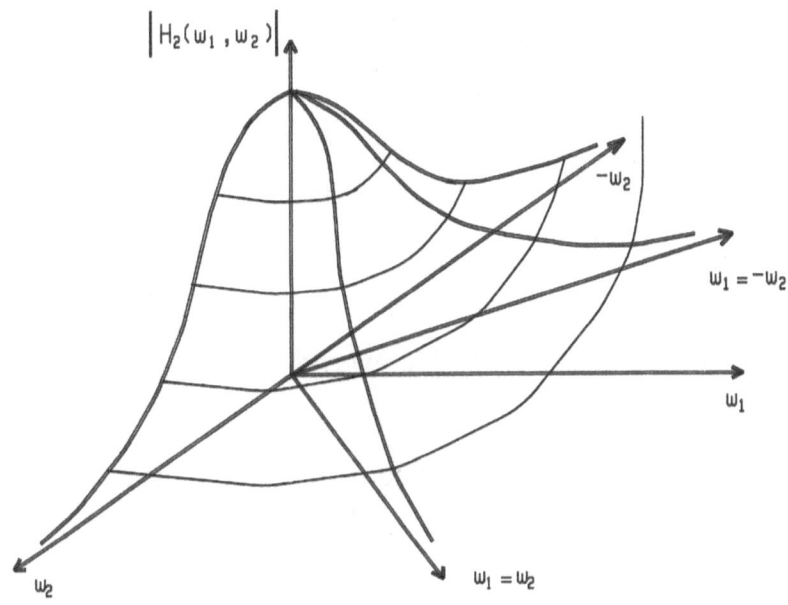

Figure 4.4 Second-order transfer function

4.5 n^{th}-ORDER VOLTERRA SYSTEMS

In this section the Volterra system is generalized to the nth-order. At this point the extension from second-order systems to higher-order ones, for all intents and purposes, is achieved by changing the subscript "2" to "n" in most of the equations developed in the previous section. Little time will be spent developing the higher-order expressions, rather the expressions from the previous section on second-order systems will simply be extended to n^{th}-order.

4.5.1 nth-ORDER IMPULSE RESPONSE

As previously stated, higher-order systems are not completely characterized by their responses to single impulses. For an nth-order Volterra system, the response to a single impulse only characterizes the system along a trajectory: $\tau_1 = \tau_2 = \cdots \tau_n$. To characterize the off-trajectory values of the impulse response, n impulses are used, and all possible combinations of the n times must, in general, be allowed for. The result of such a set of inputs is an nth-order impulse response that completely characterizes the system.

Equation 4.60 is the general equation governing an nth-order Volterra system.

$$Y_n[x(t)] = \iint_{-\infty}^{\infty} \cdots \int_{-\infty}^{\infty} \frac{h_n(\tau_1, \ldots, \tau_n) x(t - \tau_1) x(t - \tau_2) \cdots}{x(t - \tau_n) d\tau_1 d\tau_2 \ldots d\tau_n} \quad (4.60)$$

The procedure for determining the nth-order impulse response proceeds as follows.

Let: $x(t) = \delta(t - T_1) + \delta(t - T_2) + \cdots + \delta(t - T_n)$ (4.61)

Substituting $x(t)$ into (4.60) yields:

$$Y_n[x(t)] = Y_n[\delta(t - T_1) + \delta(t - T_2) + \cdots + \delta(t - T_n)] \quad (4.62)$$

$$Y_n[x(t)] = \int_{-\infty}^{\infty} \cdots \int_{-\infty}^{\infty} h_n(\tau_1, \tau_2, \ldots, \tau_n)[\delta(t - T_1 - \tau_1) +$$

$$\delta(t - T_2 - \tau_1) + \cdots + \delta(t - T_n - \tau_1)][\delta(t - T_1 - \tau_2) +$$

$$\delta(t - T_2 - \tau_2) + \cdots + \delta(t - T_n - \tau_2)] \ldots [\delta(t - T_1 - \tau_n) +$$

$$\delta(t - T_2 - \tau_n) + \cdots + \delta(t - T_n - \tau_n)]d\tau_1 d\tau_2 \ldots d\tau_n \qquad (4.63)$$

The expansion of (4.63) contains many terms. If symmetric kernels are assumed, then only one term resulting from the expansion, namely the n-linear term, completely characterizes the system. The other terms contain only redundant information. Specifically, they represent the system along certain distinct trajectories. Therefore, the nth-order impulse response for symmetric kernels is completely described by:

$$g_n(t) = h_n(t - T_1, t - T_2, \ldots, t - T_n) \qquad (4.64)$$

In other words, the symmetric n^{th}-order impulse response is the n-linear term divided by n!. The division by n! is required to eliminate the n! generated from the symmetrization process.

4.5.2 KERNEL SYMMETRIZATION

By extending the results of the second-order symmetrization process it can be shown that for an nth-order system, an equivalent symmetric kernel can be obtained from the asymmetric ones as follows.

$$h_n^{sym}(\tau_1, \tau_2, \dots, \tau_n) = \frac{1}{n!}\sum_{\substack{over\ all\ \tau \\ permutations}} h_n^{asym}(\tau_1, \tau_2, \dots, \tau_n)$$

(4.65)

There are n! different permutations of the argument $(\tau_1, \tau_2, \dots, \tau_n)$. Therefore, it follows that to produce a symmetric kernel the result of the summation needs to be divided by n!. When n=2, (4.65) reduces to (4.47).

4.5.3 CAUSALITY

Recall that a causal system is one that does not depend on the future for any input. By extension of the second-order case an nth-order system is causal if and only if:

$$h_n(\tau_1, \tau_2, \dots, \tau_n) = 0 \quad for\ \tau_i < 0,\ i = 1, 2, \dots, n$$

(4.66)

4.5.4 STABILITY

The "BIBO" stability criterion is again selected. Therefore, it can be shown that a sufficient condition for stability for an nth-order system is given by (4.67) below.

$$\int_{-\infty}^{\infty} \dots \int_{-\infty}^{\infty} |h_n(\tau_1, \tau_2, \dots, \tau_n)| d\tau_1 d\tau_2 \dots d\tau_n < 0$$

(4.67)

Note that, as in the second-order case, this is sufficient to guarantee stability, but not necessary.

4.5.5 nth-ORDER VOLTERRA TRANSFER FUNCTION

As in the first- and second-order cases, the nth-order Volterra transfer function is the Fourier transform of the nth-order Volterra kernel (n^{th}-order impulse response). Now, however, the kernel is a function of n variables and requires an n-dimensional Fourier transform. The generalization of the Fourier transform to functions of n variables is accomplished by transforming the n-dimensional function with respect to each variable one at a time as was done in the two-dimensional case. This process results in the following multidimensional transform pair.

$$H_n(\omega_1, \ldots, \omega_n) = \int_{-\infty}^{\infty} \cdots \int_{-\infty}^{\infty} h_n(\tau_1, \ldots, \tau_n) e^{-j(\omega_1\tau_1 + \cdots + \omega_n\tau_n)} \, d\tau_1 \ldots d\tau_n$$

$$(4.68)$$

$$h_n(\tau_1, \ldots \tau_n) = \frac{1}{(2\pi)^n} \int_{-\infty}^{\infty} \cdots \int_{-\infty}^{\infty} H_n(\omega_1, \ldots, \omega_n) e^{j(\omega_1\tau_1 + \cdots + \omega_n\tau_n)} \, d\omega_1 \ldots d\omega_n$$

$$(4.69)$$

The function $H_n(\omega_1, \ldots, \omega_n)$ is the nth-order Volterra transfer function and characterizes the nth-order system response in the frequency domain.

Note that $H_n(\omega_1, \ldots, \omega_n)$ is a symmetric function if $h_n(\tau_1, \ldots, \tau_n)$ is symmetric.

As was the case for the lower-order systems, (4.67) ensures the existence of the n-dimensional Fourier transform of the Volterra kernel $h_n(\tau_1, \ldots, \tau_n)$.

4.5.6 SINUSOIDAL RESPONSE OF nth-ORDER VOLTERRA SYSTEMS

Proceeding in an identical manner to the method used in the lower-order cases:

Let: $x(t) = A\cos(w_0 t) = \sum_{q=-1}^{1} \frac{A}{2} e^{jqw_0 t}$; $q \neq 0$ (3.3)

Substituting (3.3) into (4.60) yields:

$$Y_n[x(t)] = Y_n\left[\sum_{q=-1}^{1} \frac{A}{2} e^{jqw_0 t}\right]$$

$$Y_n[x(t)] = \frac{A^n}{2^n} \int_{-\infty}^{\infty} \cdots \int_{-\infty}^{\infty} h_n(\tau_1, \ldots, \tau_n)\left[\sum_{q_1=-1}^{1} e^{jq_1 w_0(t-\tau_1)}\right] \cdot$$
$$\cdots \cdot \left[\sum_{q_n=-1}^{1} e^{jq_n w_0(t-\tau_n)}\right] d\tau_1 \ldots d\tau_n \qquad (4.70)$$

Rearranging the summations and integrations and pulling out terms that are constant with respect to the integration variables τ_i, (4.70) can be rewritten as:

$$Y_n[x(t)] = \frac{A^n}{2^n} \sum_{q_1=-1}^{1} \cdots \sum_{q_n=-1}^{1} e^{j(q_1 w_0 + \cdots + q_n w_0)t}$$

$$\cdot \int_{-\infty}^{\infty} \cdots \int_{-\infty}^{\infty} h_n(\tau_1, \ldots \tau_n) e^{-j(q_1 w_0 \tau_1 + \cdots + q_n w_0 \tau_n)} \, d\tau_1 \ldots d\tau_n$$

$$(4.71)$$

Now substituting (4.68) into (4.71) yields:

$$Y_n[x(t)] = \frac{A^n}{2^n} \sum_{q_1=-1}^{1} \cdots \sum_{q_n=-1}^{1} e^{j(q_1 w_0 + \cdots + q_n w_0)t}$$

$$\cdot H_n(q_1 w_0, \ldots, q_n w_0) \qquad (4.72)$$

Equation 4.72 represents the sinusoidal response of an nth-order Volterra system. When $n = 2$, (4.72) reduces to (4.56).

Comparing (4.72) with the corresponding equation for a memoryless system, (3.8), illustrates some interesting similarities and some critical differences.

$$Y_n[x(t)] = \frac{A^n}{2^n} \sum_{q_1=-1}^{1} \cdots \sum_{q_n=-1}^{1} a_n e^{j(q_1 w_0 + \cdots + q_{nw_0})t} \qquad (3.8)$$

The obvious similarity is the general form. Both the number of harmonics and their location in the frequency domain are identical. The only difference is in the harmonic coefficients. In the power series representation the coefficient, a_n is real and frequency independent, as expected for a memoryless system, whereas the harmonic coefficient of the Volterra

73

series representation, $H_n(q_1\omega_0, ..., q_n\omega_0)$, is both complex and frequency dependent.

This difference illustrates the power of Volterra series analysis. Not only can the magnitudes of the harmonics be determined, but also their relative phases and frequency response. When power series analysis is used to analyze a system with memory, the system first needs to be approximated by a memoryless one. Sometimes this approximation is impossible to perform, and even when it is possible, errors are inevitably introduced. Depending on the type of energy storage devices as well as the type of nonlinearities, these errors can be quite significant. This inability to determine the relative phases of harmonics can lead to errors when calculations involving harmonic coefficients are required (i.e. gain compression, harmonic content, third-order intercept). As scalar quantities, power series coefficients can only add in or out of phase with each other. Therefore, errors occur when the harmonics being added are not in phase. The greater the phase difference the larger the error. The largest error occurs when the harmonics being added are thought to be in phase when in reality are out of phase with respect to each other. This limitation is further exacerbated by the lack of frequency information in the power series coefficients. Two harmonics may be in phase at one frequency but out of phase at another. This situation often results in the calculated harmonic values being either too large or too small. In either case the calculations are inaccurate.

Volterra coefficients are frequency-dependent vector quantities and as such add vectorially. Both their magnitudes and relative phases are used in any calculations. Therefore, the calculated harmonic values are more accurate and this accuracy is maintained over frequency.

Finally, if the Volterra transfer function in (4.72) is assumed to be symmetric, then combinational analysis techniques [8] can be applied to further simplify it as will be demonstrated.

If k equals the number of times $-\omega_0$ appears as an argument in the nth-order Volterra transfer function, then $n - k$ will refer to the number of times $+\omega_0$ appears and the corresponding frequency will be defined by $\omega = (n - 2k)\omega_0$.

The number of different ways $-\omega_0$ and $+\omega_0$ can be arranged to produce the frequency $(n - 2k)\omega_0$ is given by the binomial coefficient:

$$\binom{n}{k} = \frac{n!}{k!(n-k)!} \qquad (4.73)$$

If all permutations for a given frequency are assumed to be equal (i.e. a symmetric transfer function), then (4.72) can be written as:

$$Y_n[x(t)] = \frac{A^n}{2^n} \sum_{k=0}^{n} \binom{n}{k} H_{k,n-k}(\omega_0) e^{j(n-2k)\omega_0 t} \qquad (4.74)$$

where $H_{k,n-k}(\omega_0)$ equals $H_n(\omega_1, \omega_2, ..., \omega_n)$ in which the first k arguments equal $-\omega_0$ and the remaining $n - k$ arguments equal $+\omega_0$.

Equation 4.74 is the general expression for the sinusoidal response due to the nth-order term in a Volterra system when the input is given by (3.3)

Summing all the various Volterra system responses yields the total response, $y(t)$.

$$y(t) = \sum_{n=1}^{\infty} Y_n[x(t)] \qquad (4.75)$$

$$y(t) = \sum_{n=1}^{\infty} \frac{A^n}{2^n} \sum_{k=0}^{n} \binom{n}{k} H_{k,n-k}(\omega_0) e^{j(n-2k)\omega_0 t} \qquad (4.76)$$

5.0 DETERMINATION OF VOLTERRA TRANSFER FUNCTIONS

5.1 INTRODUCTION

The next step is to develop a method for quantifying the harmonic coefficients, i.e. the Volterra transfer functions. An efficient and somewhat intuitive approach for determining the Volterra transfer functions is called the "Harmonic Input" or "Probing" method [2], [6], [8], [11], [25]. This method relies on the fact that (4.1) will produce a harmonic output given a harmonic input. This is a frequency domain technique and therefore, results in the determination of the Volterra transfer function $H_n(\omega_1, \omega_2, \dots, \omega_n)$. If needed, the nonlinear impulse response could be obtained using the inverse Fourier transform. Because calculating multi-dimensional Fourier transforms is tedious however, and most of the important information about a nonlinear system is easily interpreted in the frequency domain, there is little motivation to use a time-domain technique.

5.2 THEORETICAL DEVELOPMENT

The method will be developed assuming an nth-order system. Therefore, the analysis begins with (4.60), which represents the output of a Volterra system due to the nth-order term in the Volterra series.

$$Y_n[x(t)] = \int\int_{-\infty}^{\infty} \cdots \int_{-\infty}^{\infty} \frac{h_n(\tau_1, \ldots, \tau_n)x(t - \tau_1) \cdots}{x(t - \tau_n)d\tau_1 \ldots \tau_n} \tag{4.60}$$

Next assume the input to this system, $x(t)$, is the sum of Q complex exponentials:

$$x(t) = e^{j\omega_1 t} + e^{j\omega_2 t} + \cdots + e^{j\omega_Q t}$$

$$x(t) = \sum_{i=1}^{Q} e^{j\omega_i t} \tag{5.1}$$

The frequencies $\omega_1, \omega_2, \ldots, \omega_Q$ form a linearly independent set; there is no set of rational numbers $m_1, m_2, \ldots, m_Q$ (some of which must be nonzero) such that:

$$\sum_{i=1}^{\infty} m_i \omega_i = 0 \tag{5.2}$$

Such sets are also called noncommensurate. This restriction is necessary to determine the output response to the entire input space [6]. Equation (5.1) is referred to as a harmonic input, hence the name of the technique.

Substituting (5.1) into (4.60) yields:

$$Y_n[x(t)] = Y_n\left[\sum_{i=1}^{Q} e^{j\omega_i t}\right]$$

$$Y_n[x(t)] = \int\int_{-\infty}^{\infty} \cdots \int_{-\infty}^{\infty} h_n(\tau_1, \tau_2, \ldots, \tau_n) \prod_{i=1}^{n} \sum_{q=1}^{Q} e^{j\omega_q(t - \tau_i)} d\tau_i$$

Expanding the product of the summation yields:

$$\prod_{i=1}^{n} \sum_{q=1}^{Q} e^{j\omega_q(t-\tau_i)} d\tau_i =$$
$$\sum_{q_1=1}^{Q} e^{j\omega_{q1}(t-\tau_1)} d\tau_1 \cdots \sum_{q_n=1}^{Q} e^{j\omega_{qn}(t-\tau_n)} d\tau_n$$

$$\prod_{i=1}^{n} \sum_{q=1}^{Q} e^{j\omega_q(t-\tau_i)} d\tau_i = \sum_{q_1=1}^{Q} \cdots \sum_{q_n=1}^{Q} \prod_{i=1}^{n} e^{j\omega_{qi}(t-\tau_i)} d\tau_i$$

$$\prod_{i=1}^{n} \sum_{q=1}^{Q} e^{j\omega_q(t-\tau_i)} d\tau_i =$$
$$\sum_{q_1=1}^{Q} \cdots \sum_{q_n=1}^{Q} \prod_{i=1}^{n} e^{j\omega_{qi}t} \prod_{i=1}^{n} e^{-j\omega_{qi}\tau_i} d\tau_i$$

Now interchanging the order of the summation and integration and pulling the constant product out of the integral yields (5.3).

$$Y_n[x(t)] = \sum_{q_1=1}^{Q} \cdots \sum_{q_n=1}^{Q} \prod_{i=1}^{n} e^{j\omega_{qi}t} \iint_{-\infty}^{\infty} \cdots \int_{-\infty}^{\infty} \frac{h_n(\tau_1, \ldots, \tau_n)}{\prod_{i=1}^{n} e^{-j\omega_{qi}\tau_i}} d\tau_i$$
(5.3)

Equation 5.3 can be further simplified by remembering that:

$$H_n(\omega_1, \omega_2, \ldots, \omega_n) = \iint_{-\infty}^{\infty} \cdots \int_{-\infty}^{\infty} h_n(\tau_1, \ldots, \tau_n) \prod_{i=1}^{n} e^{-j\omega_i\tau_i} d\tau_i$$
(4.68)

Therefore, substituting (4.68) into (5.3) yields:

$$Y_n[x(t)] = \sum_{q_1=1}^{Q} \cdots \sum_{q_n=1}^{Q} H_n(\omega_{q1}, \ldots, \omega_{qn}) e^{j(\omega_{q1}+\cdots+\omega_{qn})t} \quad (5.4)$$

Finally, summing the terms of all the different orders yields the total response, $y(t)$:

$$y(t) = \sum_{n=1}^{\infty} \sum_{q_1=1}^{Q} \cdots \sum_{q_n=1}^{Q} H_n(\omega_{q1}, \ldots, \omega_{qn}) e^{j(\omega_{q1}+\cdots+\omega_{qn})t}$$

(5.5)

When $Q = n$ in (5.5) there will be exactly $n!$ terms that have the exponential multiplier $e^{j(\omega_1+\cdots+\omega_n)t}$. Each term results from a different combination of $\omega_1, \ldots, \omega_n$. By requiring that $(\omega_1, \ldots, \omega_n)$ form an independent set, as was done in (5.2) no other order Volterra operator will contribute terms associated with the harmonic $e^{j(\omega_1+\cdots+\omega_n)t}$. Finally, by assuming symmetric kernels, a general expression representing the harmonic, $e^{j(\omega_1+\cdots+\omega_n)t}$, in (5.5) can be written as follows:

$$y(t)|_{@\omega_1+\cdots+\omega_n} = n! \, H_n(\omega_1, \ldots, \omega_n) e^{j(\omega_1+\cdots+\omega_n)t}$$

(5.6)

Therefore, it follows that the symmetric n^{th}-order Volterra transfer function, $H_n(\omega_1, \ldots, \omega_n)$ can be obtained analytically by dividing the coefficient of $e^{j(\omega_1+\cdots+\omega_n)t}$ in (5.6) by $n!$:

$$H_n(\omega_1, \ldots, \omega_n) = \frac{1}{n!} \cdot coefficient \; of \; e^{j(\omega_1+\cdots+\omega_n)t}$$

(5.7)

Note that this is purely an analytical method for finding Volterra transfer functions since the sum of linearly independent complex exponentials is not a real function of time.

5.3 APPLICATION

In order to clarify this method it will be applied to the following nonlinear circuit consisting of a linear capacitor, a linear resistor and a nonlinear resistor in parallel with a current source:

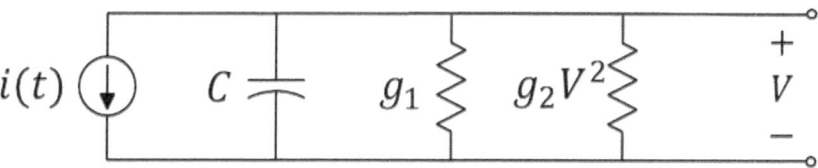

Figure 5.1 Nonlinear Circuit

The nonlinear system equation relating the input, $i(t)$, to the output, $v(t)$, is given by:

$$i(t) = C\frac{dv(t)}{dt} + g_1 v(t) + g_2 v^2(t) \tag{5.8}$$

Since Equation (5.8) is valid for any time or frequency, the method of harmonic balance [10] can be used to solve it. The method of harmonic balance can be viewed as an extension of phasor analysis for linear differential equations. When an equation is nonlinear, the steady-state solution is no longer a single sinusoid but rather a linear combination of sinusoids (i.e. a fundamental plus harmonics resulting from the nonlinearities). By requiring the sinusoids be harmonically

81

related the solution will be periodic. Substituting the assumed solution into a nonlinear differential equation results in an algebraic equation that can be factored into a sum of purely sinusoidal terms. Taking advantage of the orthogonality property of sinusoids results in a set of simpler independent algebraic equations, one for each harmonic in the assumed solution. The equations are solved by determining the sinusoidal amplitudes that satisfy or balance them at each harmonic.

DETERMINE $H_1(\omega_1)$:

1. Let $i(t) = x(t)$ and $Q = 1$ in the harmonic input equation (5.1):

$$i(t) = e^{j\omega_1 t} \qquad\qquad (5.9)$$

2. The first-order response for this input can be found by taking the first term of (5.5). The expansion is truncated at n=1 because higher-order terms do not contribute to the first-order transfer function.

$$v(t) = v_1(t) = Y_1[x(t)] = H_1(\omega_1)\, e^{j\omega_1 t} \qquad\qquad (5.10)$$

3. Substitute both $i(t)$ and $v(t)$ into the system equation given by (5.8):

$$e^{j\omega_1 t} = C \frac{d[H_1(\omega_1)e^{j\omega_1 t}]}{dt} + g_1[H_1(\omega_1)e^{j\omega_1 t}]$$
$$+ g_2[H_1(\omega_1)e^{j\omega_1 t}]^2$$

$$e^{j\omega_1 t} = j\omega_1 C H_1(\omega_1)e^{j\omega_1 t} + g_1 H_1(\omega_1)e^{j\omega_1 t}$$
$$+ g_2 H_1^2(\omega_1)e^{j2\omega_1 t} \qquad (5.11)$$

The method of harmonic balance requires that each harmonic component be individually satisfied. Therefore, equating the coefficients of $e^{j\omega_1 t}$ on both sides of (5.11) yields:

$$1 = j\omega_1 C H_1(\omega_1) + g_1 H_1(\omega_1)$$

Therefore,

$$H_1(\omega_1) = \frac{1}{g_1 + j\omega_1 C} \qquad (5.12)$$

Notice that the first-order Volterra transfer function (5.12) is the same result that would be obtained using conventional linear methods on the circuit in Figure 5.1 with $g_2 = 0$. This should not be a surprise since when $g_2 = 0$ the circuit becomes a standard linear system.

DETERMINE $H_2(\omega_1, \omega_2)$:

1. Let: $i(t) = x(t)$ and $Q = 2$ in (5.1):

$$i(t) = \sum_{i=1}^{2} e^{j\omega_i t} \qquad (5.13)$$

83

2. $v(t)$ is found by taking the first two terms of (5.5) i.e. truncate at $n = 2$. Again, higher-order terms play no role in the determination of the second-order transfer function.

$$v(t) = v_1(t) + v_2(t) \qquad (5.14)$$

$$v(t) = \sum_{n=1}^{2} \sum_{q_1=1}^{2} \cdots \sum_{q_n=1}^{2} \frac{H_n(\omega_{q1}, \ldots, \omega_{qn})}{\cdot \, e^{j(\omega_{q1}+\cdots+\omega_{qn})t}} \qquad (5.15)$$

$$n = 1 \implies v_1(t) = \sum_{q_1=1}^{2} H_1(\omega_{q1}) e^{j\omega_{q1}t} \qquad (5.16)$$

$$v_1(t) = H_1(\omega_1) e^{j\omega_1 t} + H_1(\omega_2) e^{j\omega_2 t} \qquad (5.17)$$

$$n = 2 \implies v_2(t) = \sum_{q_1=1}^{2} \sum_{q_2=1}^{2} H_2(\omega_{q1}, \omega_{q2}) e^{j(\omega_{q1}+\omega_{q2})t} \qquad (5.18)$$

$$v_2(t) = \sum_{q_1=1}^{2} \left[\begin{array}{l} H_2(\omega_{q1}, \omega_1) e^{j(\omega_{q1}+\omega_1)t} \\ +H_2(\omega_{q1}, \omega_2) e^{j(\omega_{q1}+\omega_2)t} \end{array} \right] \qquad (5.19)$$

$$v_2(t) = H_2(\omega_1, \omega_1) e^{j2\omega_1 t} + H_2(\omega_2, \omega_2) e^{j2\omega_2 t}$$
$$+H_2(\omega_1, \omega_2) e^{j(\omega_1+\omega_2)t} + H_2(\omega_2, \omega_1) e^{j(\omega_2+\omega_1)t} \qquad (5.20)$$

Substituting (5.17) and (5.20) into (5.14) yields the final result for $v(t)$:

$$v(t) = H_1(\omega_1) e^{j\omega_1 t} + H_1(\omega_2) e^{j\omega_2 t} + H_2(\omega_1, \omega_1) e^{j2\omega_1 t}$$
$$+ H_2(\omega_2, \omega_2) e^{j2\omega_2 t} + H_2(\omega_1, \omega_2) e^{j(\omega_1+\omega_2)t}$$
$$+H_2(\omega_2, \omega_1) e^{j(\omega_2+\omega_1)t} \qquad (5.21)$$

3. Substitute $i(t)$ and $v(t)$ into (5.8):

$$e^{j\omega_1 t} + e^{j\omega_2 t} =$$

$$C\frac{d}{dt}\begin{bmatrix} H_1(\omega_1)e^{j\omega_1 t} + H_1(\omega_2)e^{j\omega_2 t} \\ +H_2(\omega_1,\omega_1)e^{j2\omega_1 t} + H_2(\omega_2,\omega_2)e^{j2\omega_2 t} \\ +H_2(\omega_1,\omega_2)e^{j(\omega_1+\omega_2)t} + H_2(\omega_2,\omega_1)e^{j(\omega_2+\omega_1)t} \end{bmatrix}$$

$$+g_1\begin{bmatrix} H_1(\omega_1)e^{j\omega_1 t} + H_1(\omega_2)e^{j\omega_2 t} \\ +H_2(\omega_1,\omega_1)e^{j2\omega_1 t} + H_2(\omega_2,\omega_2)e^{j2\omega_2 t} \\ H_2(\omega_1,\omega_2)e^{j(\omega_1+\omega_2)t} + H_2(\omega_2,\omega_1)e^{j(\omega_2+\omega_1)t} \end{bmatrix}$$

$$+g_2\begin{bmatrix} H_1(\omega_1)e^{j\omega_1 t} + H_1(\omega_2)e^{j\omega_2 t} \\ +H_2(\omega_1,\omega_1)e^{j2\omega_1 t} + H_2(\omega_2,\omega_2)e^{j2\omega_2 t} \\ +H_2(\omega_1,\omega_2)e^{j(\omega_1+\omega_2)t} + H_2(\omega_2,\omega_1)e^{j(\omega_2+\omega_1)t} \end{bmatrix}^2$$

$$(5.22)$$

Now, as in the first-order case, a harmonic balance calculation is performed. In order to simplify the math, only the terms associated with the exponentials $e^{j(\omega_1+\omega_2)t}$ or $e^{j(\omega_2+\omega_1)t}$ will be considered since their coefficients will determine the second-order Volterra transfer function $H_2(\omega_1,\omega_2)$:

$$0 = j(\omega_1 + \omega_2)CH_2(\omega_1,\omega_2)e^{j(\omega_1+\omega_2)t}$$
$$+j(\omega_2 + \omega_1)CH_2(\omega_2,\omega_1)e^{j(\omega_2+\omega_1)t}$$
$$+g_1\left[H_2(\omega_1,\omega_2)e^{j(\omega_1+\omega_2)t} + H_2(\omega_2,\omega_1)e^{j(\omega_2+\omega_1)t}\right]$$
$$+g_2\left[H_1(\omega_1)H_1(\omega_2)e^{j(\omega_1+\omega_2)t} + H_1(\omega_2)H_1(\omega_1)e^{j(\omega_2+\omega_1)t}\right]$$

$$(5.23)$$

Notice that the left side of (5.23) is equal to zero. This is because the input, $i(t)$, contained no frequency components at either $(\omega_1 + \omega_2)$ or $(\omega_2 + \omega_1)$.

Equation (5.23) can be simplified further by assuming symmetric transfer functions, i.e. $H_2(\omega_1, \omega_2) = H_2(\omega_2, \omega_1)$ and dividing out the term $e^{j(\omega_1 + \omega_2)t}$.

$$0 = j2C(\omega_1 + \omega_2)H_2(\omega_1, \omega_2) + 2g_1 H_2(\omega_1, \omega_2)$$
$$+2g_2 H_1(\omega_1)H_1(\omega_2)$$

$$H_2(\omega_1, \omega_2) = \frac{-g_2 H_1(\omega_1)H_1(\omega_2)}{[g_1 + jC(\omega_1 + \omega_2)]} \tag{5.24}$$

Substituting for $H_1(\omega_1)$ and $H_1(\omega_2)$ yields the second-order Volterra transfer function entirely in terms of circuit parameters:

$$H_2(\omega_1, \omega_2) = \frac{-g_2}{(g_1 + jC\omega_1)(g_1 + jC\omega_2)[g_1 + jC(\omega_1 + \omega_2)]} \tag{5.25}$$

Notice that (5.25) could also be written in terms of the lower-order Volterra transfer function, namely $H_1(\cdot)$:

$$H_2(\omega_1, \omega_2) = -g_2 H_1(w_1)H_1(w_2)H_1(\omega_1 + \omega_2) \tag{5.26}$$

DETERMINE $H_3(\omega_1, \omega_2, \omega_3)$:

1. Let $i(t) = x(t)$ and $Q = 3$ in (5.1):

$$i(t) = \sum_{i=1}^{3} e^{j\omega_i t}$$
(5.27)

2. Take the first three terms of (5.5), i.e. truncate at $n = 3$.

$$v(t) = v_1(t) + v_2(t) + v_3(t)$$
(5.28)

$$v(t) = \sum_{n=1}^{3} \sum_{q_1=1}^{3} \cdots \sum_{q_n=1}^{3} \frac{H_n(\omega_{q1}, \ldots, \omega_{qn})}{\cdot e^{j(\omega_{q1} + \cdots + \omega_{qn})t}}$$
(5.29)

$$n = 1 \implies v_1(t) = \sum_{q_1=1}^{3} H_1(\omega_{q1}) e^{j\omega_{q1} t}$$
(5.30)

$$n = 2 \implies v_2(t) = \sum_{q_1=1}^{3} \sum_{q_2=1}^{3} \frac{H_2(\omega_{q1}, \omega_{q2})}{\cdot e^{j(\omega_{q1} + \omega_{q2})t}}$$
(5.31)

$$n = 3 \implies v_3(t) = \sum_{q_1=1}^{3} \sum_{q_2=1}^{3} \sum_{q_3=1}^{3} \frac{H_3(\omega_{q1}, \omega_{q2}, \omega_{q3})}{\cdot e^{j(\omega_{q1} + \omega_{q2} + \omega_{q3})t}}$$
(5.32)

3. Substitute the expanded forms of $i(t)$ and $v(t)$, (5.27) and (5.29) respectively, into the system equation (5.8) and balance the $e^{j(\omega_1 + \omega_2 + \omega_3)t}$ harmonic coefficients assuming symmetry:

87

$$0 = j6C(\omega_1 + \omega_2 + \omega_3)H_3(\omega_1, \omega_2, \omega_3)$$
$$+ 6g_1 H_3(\omega_1, \omega_2, \omega_3) + 4g_2 H_2(\omega_1, \omega_2)H_1(\omega_3)$$
$$+ 4g_2 H_2(\omega_1, \omega_3)H_1(\omega_2) + 4g_2 H_2(\omega_2, \omega_3)H_1(\omega_1)$$

$$0 = 6H_3(\omega_1, \omega_2, \omega_3)[g_1 + jC(\omega_1 + \omega_2 + \omega_3)]$$
$$+ 4g_2 [H_1(\omega_1)H_2(\omega_2, \omega_3) + H_1(\omega_2)H_2(\omega_1, \omega_3)$$
$$+ H_1(\omega_3)H_2(\omega_1, \omega_2)]$$

$$H_3(\omega_1, \omega_2, \omega_3) =$$
$$\frac{-2g_2[H_1(\omega_1)H_2(\omega_2,\omega_3)+H_1(\omega_2)H_2(\omega_1,\omega_3)+H_1(\omega_3)H_2(\omega_1,\omega_2)]}{3[g_1+jC(\omega_1+\omega_2+\omega_3)]} \qquad (5.33)$$

Equation (5.33) can now be rewritten in terms of $H_1(\cdot)$ and $H_2(\cdot)$ as follows:

$$H_3(\omega_1, \omega_2, \omega_3) =$$
$$-\frac{2}{3}g_2 H_1(\omega_1 + \omega_2 + \omega_3)\begin{bmatrix} H_1(\omega_1)H_2(\omega_2, \omega_3) \\ +H_1(\omega_2)H_2(\omega_1, \omega_3) \\ +H_1(\omega_3)H_2(\omega_1, \omega_2) \end{bmatrix} \qquad (5.34)$$

Note that (5.33) can also be rewritten entirely in terms of the first-order Volterra transfer function, $H_1(\cdot)$:

$$H_3(\omega_1, \omega_2, \omega_3) = \frac{2}{3}g_2^2 H_1(\omega_1)H_1(\omega_2)H_1(\omega_3)$$
$$\cdot \begin{bmatrix} H_1(\omega_1 + \omega_2) \\ +H_1(\omega_1 + \omega_3) \\ +H_1(\omega_2 + \omega_3) \end{bmatrix} H_1(\omega_1 + \omega_2 + \omega_3)$$

$$(5.35)$$

If desired (5.12) and (5.25) can be substituted into (5.33) to yield the third-order Volterra transfer function exclusively in terms of circuit elements.

This problem illustrates some interesting points worth noting:

1. A third-order Volterra transfer function exists for the given circuit even though the nonlinearity was limited to second degree. In general, there will exist an infinite number of Volterra transfer functions for a given nonlinearity.

2. The higher the order, the more tedious and cumbersome the calculation. Thereby illustrating the desire for limiting the required Volterra transfer function to third order. This is only possible for weakly nonlinear systems.

3. Any Volterra transfer function can be written in terms of lower order ones, ultimately in terms of the linear transfer function if desired. This fact has important consequences with regard to convergence and will be discussed later.

4. The ability to represent the Volterra transfer functions in terms of circuit elements provides much more insight into the operation of the network than some of the other nonlinear analysis methods, such as harmonic balance and piecewise linear techniques.

6.0 METHOD OF NONLINEAR CURRENTS

6.1 INTRODUCTION

In the previous chapter, a Volterra transfer function was found by simply applying the harmonic input method to the nonlinear differential equation describing a circuit. This method becomes increasingly impractical for large circuits containing many nodes. In order to lay the foundation for analyzing more complex systems, the method of nonlinear currents must be developed [2], [6], [11], [24], [28]. This method has the advantage of allowing the determination of Volterra transfer functions from the solutions of linear differential equations. This feature is very powerful because it permits the use of standard linear analysis tools. The equations needed are derived from linearized versions of the original nonlinear equations and are referred to as associated linear differential equations. Their current excitations are related to the system nonlinearities and hence are known as the nonlinear currents.

This ability to use linear analysis can greatly simplify very complex nonlinear systems.

6.2 THEORETICAL DEVELOPMENT

The development of this analysis method begins with (4.1), which represents the general response of a Volterra system.

$$y(t) = \sum_{n=1}^{\infty} \int_{-\infty}^{\infty} \cdots \int_{-\infty}^{\infty} \frac{h_n(\tau_1, \ldots, \tau_n)x(t - \tau_1) \cdots}{x(t - \tau_n)d\tau_1 \ldots d\tau_n} \qquad (4.1)$$

Let: $v(t) = y(t)$ and $x(t) = z \cdot i(t)$, where z is a dummy variable used to indicate the order of the response.

$$v(t) = \sum_{n=1}^{\infty} z^n \int_{-\infty}^{\infty} \cdots \int_{-\infty}^{\infty} \frac{h_n(\tau_1, \ldots, \tau_n)i(t - \tau_1) \cdots}{i(t - \tau_n)d\tau_1 \ldots d\tau_n} \qquad (6.1)$$

Rewriting (6.1) in a more compact form yields:

$$v(t) = \sum_{n=1}^{\infty} z^n v_n[i(t)] \qquad (6.2)$$

Equation (6.2) states that the output voltage of a nonlinear system, using Volterra series analysis, is an infinite series of voltages each representing a different order response.

Note that the input could just as easily have been a voltage and the output a current, which would yield:

$$i(t) = \sum_{n=1}^{\infty} z^n i_n[v(t)] \qquad (6.3)$$

Equations (6.2) and (6.3) demonstrate that for a Volterra system, (i.e. a nonlinear system that can be represented by a convergent Volterra series), the node voltages and branch currents can be expressed in a power series in z in which the coefficient of the z^n term represents the nth-order components. The application of Kirchhoff's current and voltage laws requires that the sum of all currents entering a node is zero and the sum of all voltages around a closed path is also zero. Therefore, since a power series equals zero if and only if each coefficient equals zero, to obey Kirchhoff's laws, the sum of all the nth-order current components entering a node and the sum of all the nth-order voltage components around a closed path must equal zero for each value of n. This result can be used to construct an equivalent linear circuit whose branch currents are represented by independent current sources that are functions of lower-order node voltages. This method results in the ability to completely analyze nonlinear networks by sequentially analyzing linear ones.

To develop this method, the response of a nonlinear voltage-controlled current source is first analyzed and the corresponding equivalent circuit constructed. The current source is described by:

$$i = \sum_{k=1}^{\infty} g_k v^k \tag{6.4}$$

Where the notation has been simplified by dropping the functional form and assuming time dependency (i.e. $i = i[v(t)]$ and $v = v[i(t)]$).

If the control voltage appears across the nonlinearity, (6.4) represents a nonlinear conductance. If the control voltage is located elsewhere in the circuit, (6.4) represents a nonlinear transconductance.

Expanding the control voltage in a power series (6.2) and substituting into (6.4) results in:

$$i = \sum_{k=1}^{\infty} g_k \left[\sum_{n=1}^{\infty} z^n v_n\right]^k \tag{6.5}$$

Further expanding (6.5) and collecting like powers of z yields:

$$i = [g_1 v_1]z + [g_1 v_2 + g_2 v_1^2]z^2 + [g_1 v_3 + g_2 2 v_1 v_2 + g_3 v_1^3]z^3$$
$$+[g_1 v_4 + g_2(2 v_1 v_3 + v_2^2) + g_3 3 v_1^2 v_2 + g_4 v_1^4]z^4 + \cdots \tag{6.6}$$

Equation (6.6) represents the total current through the nonlinear element.

Substituting (6.3) into (6.6) and equating like powers of z results in the following:

$$i_1 = g_1 v_1 \tag{6.7}$$

$$i_2 = g_1 v_2 + g_2 v_1^2 \tag{6.8}$$

$$i_3 = g_1 v_3 + g_2 2 v_1 v_2 + g_3 v_1^3 \tag{6.9}$$

$$i_4 = g_1 v_4 + g_2(2 v_1 v_3 + v_2^2) + g_3 3 v_1^2 v_2 + g_4 v_1^4 \tag{6.10}$$

.   .
.   .
.   .

Equation (6.7) represents the linear or first-order portion of the overall nonlinear conductance and can be modeled as shown in Figure 6.1.

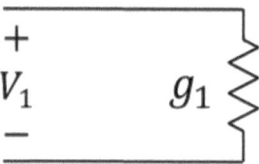

Figure 6.1 First-order Model

Equation (6.8) represents the relationship between the second-order current, i_2, and voltage, v_2, and can be modeled by the Norton equivalent circuit shown in Figure 6.2.

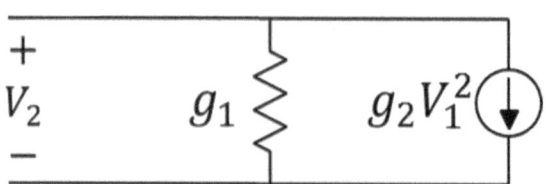

Figure 6.2 Second-order Model

Note that v_2 is the voltage across the same linear resistor as in Figure 6.1 but in parallel with a current source. The control voltage v_1 is fictitious. It only appears if the first-order term is used alone as in Figure 6.1.

Likewise the third- and fourth-order current-voltage relationships can be modeled by the circuits shown in Figure 6.3 and 6.4 respectively.

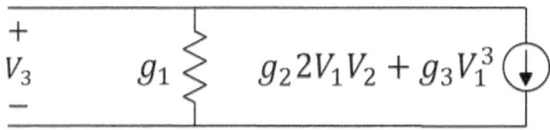

Figure 6.3 Third-order Model

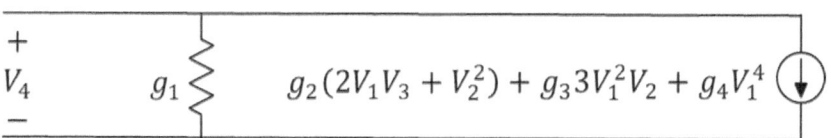

Figure 6.4 Fourth-order Model

Figure 6.2 shows that the second-order voltage response, v_2, is the response of the linearized portion of the network driven by the current source $g_2 v_1^2$. This current is referred to as the second-order nonlinear current, i_{NL2}. Similarly, Figures 6.3 and 6.4 suggest that the higher-order responses $v_3, v_4, \dots, v_n$ can be represented by the response of the linear circuit when driven by the appropriate nonlinear current sources $i_{NL2}, i_{NL3}, i_{NL4} \dots, i_{NLn}$ respectively. Therefore, for a nonlinear conductance, the nonlinear currents are:

$$i_{NL2} = g_2 v_1^2 \qquad (6.11)$$

$$i_{NL3} = g_2 2v_1 v_2 + g_3 v_1^3 \qquad (6.12)$$

$$i_{NL4} = g_2(2v_1 v_3 + v_2^2) + g_3 3v_1^2 v_2 + g_4 v_1^4 \qquad (6.13)$$

The total nonlinear current is then given by:

$$i_{NL}(t) = \sum_{n=2}^{\infty} i_{NLn}(t) \qquad (6.14)$$

An equivalent procedure can be applied to a nonlinear capacitance:

$$C = C_1 + C_2 v + C_3 v^2 + \cdots$$

$$C = \sum_{k=1}^{\infty} C_k v^{k-1} \qquad (6.15)$$

$$i = \frac{dQ}{dt} = \frac{d[C \cdot v]}{dt} = \sum_{k=1}^{\infty} C_k \frac{d[v]^k}{dt} \qquad (6.16)$$

Again representing the total voltage by (6.2) and substituting into (6.16) yields:

$$i = \sum_{k=1}^{\infty} C_k \frac{d}{dt} [\sum_{n=1}^{\infty} z^n \cdot v_n]^k \qquad (6.17)$$

Substituting (6.3) into (6.17), expanding both sides of the equation and equating like powers of z results in the following:

$$i_1 = C_1 \frac{dv_1}{dt} \tag{6.18}$$

$$i_2 = C_1 \frac{dv_2}{dt} + C_2 \frac{d[v_1^2]}{dt} \tag{6.19}$$

$$i_3 = C_1 \frac{dv_3}{dt} + C_2 \frac{d[2v_1 v_2]}{dt} + C_3 \frac{d[v_1^3]}{dt} \tag{6.20}$$

$$.$$
$$.$$
$$.$$

Therefore, the nonlinear currents are:

$$i_{NL2} = C_2 \frac{d[v_1^2]}{dt} \tag{6.21}$$

$$i_{NL3} = C_2 \frac{d[2v_1 v_2]}{dt} + C_3 \frac{d[v_1^3]}{dt} \tag{6.22}$$

Modeling (6.18), (6.19) and (6.20) in circuit form yields:

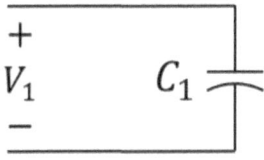

Figure 6.5 First-order Model

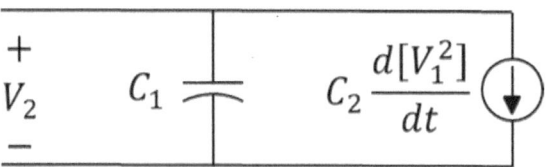

Figure 6.6 Second-order Model

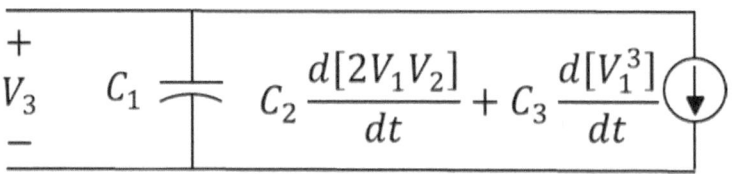

Figure 6.7 Third-order Model

6.3 FREQUENCY-DOMAIN REPRESENTATION OF NONLINEAR CURRENTS

Since the ultimate goal has been to compute the Volterra transfer functions, the method of nonlinear currents must be expanded to allow their determination. To facilitate this, it is necessary to establish what the nonlinear currents are when the excitation is a sum of distinct complex exponentials (5.1) as used in the harmonic input method. Recall that this operation results in the frequency-domain representation of the original time-domain quantities. Note this operation can also be viewed as taking the Fourier transform of the nonlinear currents. As in the previous chapter, upper-case letters are used to denote frequency-domain functions and lower-case letters are used for time-domain functions.

Therefore, i_{NL} and I_{NL} form Fourier transform pairs. As in the previous chapter we use an input of the form:

$$x(t) = \sum_{i=1}^{Q} e^{j\omega_i t} \tag{5.1}$$

Remember as developed in Chapter 5, the frequencies must be linearly independent and Q is equal to the order of the transfer function being determined.

For an input given by (5.1), the output of a Volterra system can be expressed by (5.5):

$$y(t) = \sum_{n=1}^{\infty} \sum_{q_1=1}^{Q} \cdots \sum_{q_n=1}^{Q} H_n\left(\omega_{q1}, \ldots, \omega_{qn}\right) e^{j(\omega_{q1} + \cdots + \omega_{qn})t} \tag{5.5}$$

For a first-order system, with a current input of the form $x(t)$ in (5.1) and $n = 1$ and $Q = 1$, the output voltage is given by:

$$v_1(t) = y_1 = \sum_{q=1}^{1} H_1(\omega_q) e^{j\omega_q t} \tag{5.10}$$

For a second-order system with $n = 2 \; and \; Q = 2$:

$$v_2(t) = y_2 = \sum_{q_1=1}^{2} \sum_{q_2=1}^{2} H_2(\omega_{q1}, \omega_{q2}) e^{j(\omega_{q1} + \omega_{q2})t} \tag{5.18}$$

For a third-order system with $n = 3 \; and \; Q = 3$:

$$v_3(t) = y_3 = \sum_{q_1=1}^{3} \sum_{q_2=1}^{3} \sum_{q_3=1}^{3} \frac{H_3(\omega_{q1}, \omega_{q2}, \omega_{q3})}{\cdot \; e^{j(\omega_{q1} + \omega_{q2} + \omega_{q3})t}} \tag{5.32}$$

In general, for an nth-order system, (5.5) is substituted into the appropriate time-domain expression (i.e. 6.11-6.13) for the nonlinear currents with $Q = n$. Upon expansion, a large spectrum of nonlinear current components is generated. Recall from the previous chapter that when the nth-order Volterra transfer function is being determined, the only nonlinear current component of interest is the one at the frequency $\omega_1 + \omega_2 + \cdots + \omega_n$. The frequency-domain expression for the nth-order nonlinear current is the coefficient of this spectral component divided by $n!$. The $n!$ comes from the assumption of symmetry.

To illustrate the procedure, the frequency-domain representation of the nonlinear currents found for the nonlinear conductance (6.11) and (6.12) will be determined:

DETERMINATION OF SECOND-ORDER NONLINEAR CURRENT, I_{NL2}:

For this second-order system, $Q = 2$ and

$$i_{NL2} = g_2 v_1^2 \tag{6.11}$$

$$i_{NL2} = g_2 [\textstyle\sum_{q=1}^{2} H_1(\omega_q) e^{j\omega_q t}]^2 \tag{6.23}$$

$$i_{NL2} = g_2 \textstyle\sum_{q1=1}^{2} \sum_{q2=1}^{2} H_1(\omega_{q1}) H_1(\omega_{q2}) e^{j(\omega_{q1}+\omega_{q2})t} \tag{6.24}$$

$$i_{NL2} = g_2 \begin{bmatrix} H_1(\omega_1)H_1(\omega_1)e^{j2\omega_1 t} \\ +2H_1(\omega_1)H_1(\omega_2)e^{j(\omega_1+\omega_2)t} \\ +H_1(\omega_2)H_1(\omega_2)e^{j2\omega_2 t} \end{bmatrix} \qquad (6.25)$$

Therefore,

$$I_{NL2}(\omega_1, \omega_2) = g_2 H_1(\omega_1)H_1(\omega_2) \qquad (6.26)$$

Equation (6.26) shows that the frequency-domain representation of the nonlinear current can be determined from the linear transfer function and the power series coefficient g_2. Note $H_1(\omega_1)$ and $H_1(\omega_2)$ represent the linear currents due to the inputs $e^{j\omega_1 t}$ and $e^{j\omega_2 t}$ respectively.

DETERMINATION OF THIRD-ORDER NONLINEAR CURRENT, I_{NL3}:

For this third-order system, $Q = 3$ and

$$i_{NL3} = g_2 2v_1 v_2 + g_3 v_1^3 \qquad (6.12)$$

$$i_{NL3} =$$
$$2g_2 \left[\sum_{q=1}^3 H_1(\omega_q)e^{j\omega_q t} \sum_{q_1=1}^3 \sum_{q_2=1}^3 H_2(\omega_{q1}, \omega_{q2})e^{j(\omega_{q1}+\omega_{q2})t} \right]$$
$$+ g_3 \left[\sum_{q=1}^3 H_1(\omega_q) e^{j\omega_q t} \right]^3 \qquad (6.27)$$

After expanding (6.27), collecting the coefficients of $e^{j(\omega_1+\omega_2+\omega_3)t}$ and dividing by 3!, the result for the frequency-domain representation of the third-order nonlinear current can be expressed as:

$$I_{NL3}(\omega_1, \omega_2, \omega_3) = \frac{2}{3} g_2 \begin{bmatrix} H_1(\omega_1)H_2(\omega_2, \omega_3) \\ +H_1(\omega_2)H_2(\omega_1, \omega_3) \\ +H_1(\omega_3)H_2(\omega_1, \omega_2) \end{bmatrix} \qquad (6.28)$$
$$+g_3 H_1(\omega_1)H_1(\omega_2)H_1(\omega_3)$$

In a completely analogous manner, the frequency-domain representations of the nonlinear currents can be determined for the nonlinear capacitor described by (6.21) and (6.22).

DETERMINATION OF I_{NL2}:

For this second-order system, Q=2 and

$$i_{NL2} = C_2 \frac{d[v_1^2]}{dt} \qquad (6.21)$$

$$i_{NL2} = C_2 \frac{d}{dt} [\Sigma_{q=1}^2 H_1(\omega_q)e^{j\omega_q t}]^2 \qquad (6.29)$$

$$i_{NL2} = C_2 \frac{d}{dt} \begin{bmatrix} H_1(\omega_1)H_1(\omega_1)e^{j2\omega_1 t} \\ + 2H_1(\omega_1)H_1(\omega_2)e^{j(\omega_1+\omega_2)t} \\ + H_1(\omega_2)H_1(\omega_2)e^{j2\omega_2 t} \end{bmatrix} \qquad (6.30)$$

$$i_{NL2} = j2\omega_1 C_2 H_1(\omega_1)H_1(\omega_1)e^{j2\omega_1 t}$$
$$+j2(\omega_1+\omega_2)C_2 H_1(\omega_1)H_1(\omega_2)e^{j(\omega_1+\omega_2)t}$$
$$+j2\omega_2 C_2 H_1(\omega_2)H_1(\omega_2)e^{j2\omega_2 t} \qquad (6.31)$$

Therefore,

$$I_{NL2}(\omega_1, \omega_2) = j(\omega_1 + \omega_2)C_2 H_1(\omega_1)H_1(\omega_2) \tag{6.32}$$

DETERMINATION OF I_{NL3}:

For this third-order system, $Q = 3$ and

$$i_{NL3} = C_2 \frac{d[2v_1 v_2]}{dt} + C_3 \frac{d[v_1^3]}{dt} \tag{6.22}$$

$i_{NL3} =$

$$2C_2 \frac{d}{dt} \left[\sum_{q=1}^{3} H_1(\omega_q) e^{j\omega_q t} \sum_{q_1=1}^{3} \sum_{q_2=1}^{3} H_2(\omega_{q1}, \omega_{q2}) e^{j(\omega_{q1}+\omega_{q2})t} \right]$$

$$+C_3 \frac{d}{dt} \left[\sum_{q=1}^{3} H_1(\omega_q) e^{j\omega_q t} \right]^3 \tag{6.33}$$

Following the same procedure used for the nonlinear conductance yields:

$$I_{NL3}(\omega_1, \omega_2, \omega_3) = \frac{2}{3} C_2 j(\omega_1 + \omega_2 + \omega_3) \begin{bmatrix} H_1(\omega_1)H_2(\omega_2, \omega_3) \\ + H_1(\omega_2)H_2(\omega_1, \omega_3) \\ + H_1(\omega_3)H_2(\omega_1, \omega_2) \end{bmatrix}$$

$$+ j(\omega_1 + \omega_2 + \omega_3)C_3 H_1(\omega_1)H_1(\omega_2)H_1(\omega_3) \tag{6.34}$$

6.4 APPLICATION TO NONLINEAR SYSTEM

To illustrate the method of nonlinear currents, the circuit in Figure 5.1 will be analyzed.

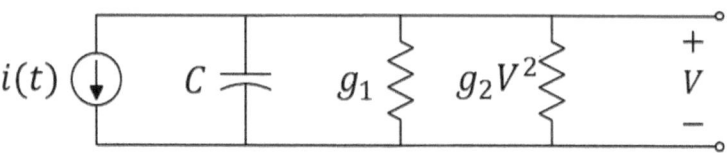

Figure 5.1 Nonlinear Circuit

The differential equation describing the circuit can be written as follows:

$$i = C\frac{dv}{dt} + \sum_{k=1}^{2} g_k v^k \tag{6.35}$$

1. Let $i = z \cdot i$ and $v = \sum_{n=1}^{\infty} z^n v_n$

$$z \cdot i = C\frac{d}{dt}\left[\sum_{n=1}^{\infty} z^n v_n\right] + \sum_{k=1}^{2} g_k \left[\sum_{n=1}^{\infty} z^n v_n\right]^k \tag{6.36}$$

2. Expanding (6.36) and equating like powers of z:

$$z \Rightarrow i = C\frac{dv_1}{dt} + g_1 v_1 \tag{6.37}$$

$$z^2 \Rightarrow 0 = C\frac{dv_2}{dt} + g_1 v_2 + g_2 v_1^2 \tag{6.38}$$

$$z^3 \Rightarrow 0 = C\frac{dv_3}{dt} + g_1 v_3 + g_2 2 v_1 v_2 \tag{6.39}$$

.   .

.   .

.   .

The nonlinearity is assumed to be mild, allowing the expansion to be terminated at $n = 3$. From (6.38) and (6.39) the nonlinear currents for the circuit in Figure 5.1 can be expressed as:

$$i_{NL2} = g_2 v_1^2 \tag{6.40}$$

$$i_{NL3} = g_2 2 v_1 v_2 \tag{6.41}$$

Equations (6.37), (6.38) and (6.39) can be written in a more demonstrative form.

$$i = \left[C \frac{d}{dt} + g_1 \right] v_1 \tag{6.42}$$

$$-i_{NL2} = -g_2 v_1^2 = \left[C \frac{d}{dt} + g_1 \right] v_2 \tag{6.43}$$

$$-i_{NL3} = -g_2 2 v_1 v_2 = \left[C \frac{d}{dt} + g_1 \right] v_3 \tag{6.44}$$

The right-hand-side of each equation represents an ordinary linear differential equation. The left-hand-side represents the nonlinearity in the form of current excitations, hence the name "nonlinear currents".

An obvious feature of the nonlinear current method is the recursive nature of the analysis, shown in steps 3 through 7 of this example.

3. Solve (6.42) for the first-order response, v_1.

4. Compute the second-order nonlinear current i_{NL2}:

$$i_{NL2} = g_2 v_1^2 \tag{6.40}$$

5. Solve (6.43) for the second-order response, v_2.

6. Compute the third-order nonlinear current i_{NL3}:

$$i_{NL3} = g_2 2 v_1 v_2 \tag{6.41}$$

7. Solve (6.44) for the third-order response, v_3.

8. Finally sum all the responses to determine the total response $v(t)$:

$$v(t) \approx \sum_{n=1}^{3} v_n(t) \tag{6.45}$$

In summary, instead of solving a nonlinear differential equation for the total response, associated linear differential equations with appropriate excitations are sequentially solved for the individual components of the nonlinear response. Recall that all the nonlinearities are lumped into the excitations. Once all the desired components (both linear and nonlinear) have been determined, the total response can be found by applying superposition. Superposition is valid even though the overall network is nonlinear since all the nonlinear responses were determined via a linear differential equation.

Once all the appropriate nonlinear currents in the time-domain have been determined, the next step is to determine their frequency-domain representation.

DETERMINE $I_{NL2}(\omega_1, \omega_2)$:

Since the second-order time-domain current for this system (6.40) is identical to (6.11) from the previous section, the results from that section can be applied here.

$$I_{NL2}(\omega_1, \omega_2) = g_2 H_1(\omega_1) H_1(\omega_2) \qquad (6.26)$$

DETERMINE $I_{NL3}(\omega_1, \omega_2, \omega_3)$:

Observe that the third-order time-domain current for this system (6.41) is very similar to (6.12) from the previous section. The only difference is that (6.12) has an additional expression, $g_3 v_1^3$. Therefore, by eliminating the terms in (6.28) that result from this expression, the frequency-domain representation of the third-order nonlinear current can be written down by inspection.

$$I_{NL3}(\omega_1, \omega_2, \omega_3) = \frac{2}{3} g_2 \begin{bmatrix} H_1(\omega_1) H_2(\omega_2, \omega_3) \\ +H_1(\omega_2) H_2(\omega_1, \omega_3) \\ +H_1(\omega_3) H_2(\omega_1, \omega_2) \end{bmatrix} \qquad (6.46)$$

At this point the analysis of the circuit in Figure 5.1 can be completed and the Volterra transfer functions determined.

DETERMINE $H_1(\omega_1)$:

$$i = \left[C\frac{d}{dt} + g_1\right]v_1 \tag{6.42}$$

$$\sum_{i=1}^{1} e^{j\omega_i t} = \left[C\frac{d}{dt} + g_1\right]\left[\sum_{q=1}^{1} H_1(\omega_q)e^{j\omega_q t}\right] \tag{6.47}$$

$$e^{j\omega_1 t} = \left[C\frac{d}{dt} + g_1\right]H_1(\omega_1)e^{j\omega_1 t} \tag{6.48}$$

$$1 = (jC\omega_1 + g_1)H_1(\omega_1) \tag{6.49}$$

$$H_1(\omega_1) = \frac{1}{(jC\omega_1 + g_1)} \tag{6.50}$$

DETERMINE $H_2(\omega_1, \omega_2)$:

$$-g_2 v_1^2 = \left[C\frac{d}{dt} + g_1\right]v_2 \tag{6.43}$$

Substitute the second-order response, (5.18) and the frequency domain representation of the second-order nonlinear current, (6.26) into (6.43).

$$-g_2 H_1(\omega_1)H_1(\omega_2) = \frac{1}{2!}\cdot coefficient\ of\ e^{j(\omega_1+\omega_2)t}\ in$$

$$\left[C\frac{d}{dt} + g_1\right]\left[\sum_{q_1=1}^{2}\sum_{q_2=1}^{2} H_2(\omega_{q_1}, \omega_{q_2})e^{j(\omega_{q_1}+\omega_{q_2})t}\right] \tag{6.51}$$

Expanding the right-hand side of (6.51) and applying harmonic balance yields (6.52). Recall that both sides of the equation represent the coefficient of the harmonic $e^{j(\omega_1+\omega_2)t}$ divided by 2!.

$$-g_2 H_1(\omega_1)H_1(\omega_2) = [jC(\omega_1 + \omega_2) + g_1]H_2(\omega_1, \omega_2) \qquad (6.52)$$

$$H_2(\omega_1, \omega_2) = \frac{-g_2 H_1(\omega_1)H_1(\omega_2)}{[jC(\omega_1+\omega_2)+g_1]} \qquad (6.53)$$

$$H_2(\omega_1, \omega_2) = -g_2 H_1(\omega_1)H_1(\omega_2)H_1(\omega_1 + \omega_2) \qquad (6.54)$$

DETERMINE $H_3(\omega_1, \omega_2, \omega_3)$:

$$-g_2 2v_1 v_2 = \left[C\frac{d}{dt} + g_1\right]v_3 \qquad (6.44)$$

Proceeding in a similar manner to the second-order case, the third-order response, (5.32) and the frequency domain representation of the third-order nonlinear current, (6.46) are substituted into (6.44).

$$-\frac{2}{3}g_2 \begin{bmatrix} H_1(\omega_1)H_2(\omega_2, \omega_3) \\ + H_1(\omega_2)H_2(\omega_1, \omega_3) \\ + H_1(\omega_3)H_2(\omega_1, \omega_2) \end{bmatrix}$$
$$= \frac{1}{3!} \cdot coefficient \ of \ e^{j(\omega_1+\omega_2+\omega_3)t} \ in \ \left[C\frac{d}{dt} + g_1\right]$$
$$\cdot \left[\sum_{q_1=1}^{3} \sum_{q_2=1}^{3} \sum_{q_3=1}^{3} H_3(\omega_{q_1}, \omega_{q_2}, \omega_{q_3})e^{j(\omega_{q_1}+\omega_{q_2}+\omega_{q_3})t}\right]$$

$$(6.55)$$

The right hand side of (6.55) is found by expanding the summations and taking $1/3!$ times the coefficient of the harmonic $e^{j(\omega_1+\omega_2+\omega_3)t}$.

$$-\frac{2}{3}g_2\begin{bmatrix} H_1(\omega_1)H_2(\omega_2,\omega_3) \\ +H_1(\omega_2)H_2(\omega_1,\omega_3) \\ +H_1(\omega_3)H_2(\omega_1,\omega_2) \end{bmatrix} = [jC(\omega_1+\omega_2+\omega_3)+g_1]H_3(\omega_1,\omega_2,\omega_3)$$

(6.56)

$$H_3(\omega_1,\omega_2,\omega_3) = \frac{-2g_2\begin{bmatrix} H_1(\omega_1)H_2(\omega_2,\omega_3) \\ +H_1(\omega_2)H_2(\omega_1,\omega_3) \\ +H_1(\omega_3)H_2(\omega_1,\omega_2) \end{bmatrix}}{3[jC(\omega_1+\omega_2+\omega_3)+g_1]}$$

(6.57)

$$H_3(\omega_1,\omega_2,\omega_3) = \frac{-2}{3}g_2\begin{bmatrix} H_1(\omega_1)H_2(\omega_2,\omega_3) \\ +H_1(\omega_2)H_2(\omega_1,\omega_3) \\ +H_1(\omega_3)H_2(\omega_1,\omega_2) \end{bmatrix} \cdot H_1(\omega_1+\omega_2+\omega_3)$$

(6.58)

All three Volterra transfer functions derived by the method of nonlinear currents; (6.50), (6.54) and (6.58), agree with the transfer functions derived in the previous chapter; (5.12), (5.25) and (5.34).

7.0 NETWORK ANALYSIS

7.1 INTRODUCTION

This chapter demonstrates the true power of the method of nonlinear currents by analyzing networks with multiple nodes.

The determination of Volterra transfer functions becomes prohibitively difficult as the number of nodes and nonlinearities increase. One way to mitigate this increase in complexity is to combine the harmonic input and nonlinear current methods with linear system theory [6], [11], [28]. Because no new concepts are introduced, this method will be illustrated by outlining the general procedure and applying it to a real circuit encountered in communication systems. Refer to Figure 7.1a-f for illustration of the procedure.

7.2 METHODOLOGY

1. Identify all the nonlinear elements in the system and the ports, called nonlinear ports, to which they are attached. The nonlinear elements are then conceptually removed from the system leaving a linear network called the "associated linear network". The nonlinear elements are then connected to their respective ports (see Figure 7.1a and 7.1b). Note the associated linear network contains

both the linear elements of the system as well as the linear portion (first-order term) of the nonlinear elements.

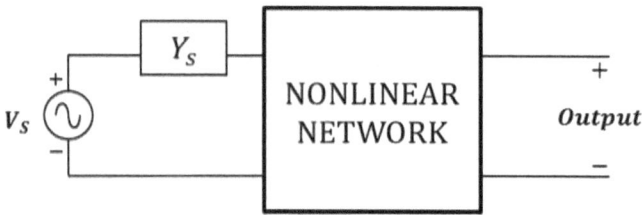

Figure 7.1a Nonlinear Network

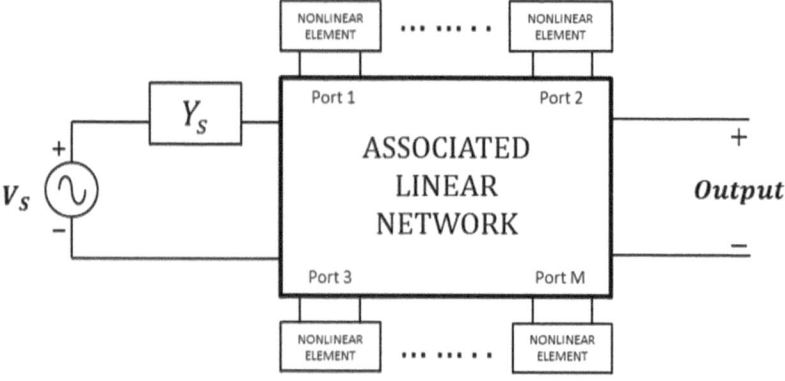

Figure 7.1b Conceptualization of Nonlinear Network with M nonlinear elements pulled out

2. Identify and label the independent nodes of the associated linear network. For this example assume there are K independent nodes.

3. Determine the nonlinear currents associated with each nonlinear element using the method described in Chapter 6.

4. Using the harmonic input method as described in Chapter 6 determine the frequency-domain representation of the nonlinear currents.

5. Replace each nonlinear element with its Norton equivalent; a linear component in parallel with the frequency-domain representation of the nonlinear current sources (see Figure 7.1c).

6. At this point, using nodal analysis, the equations governing the nonlinear system shown in Figure 7.1c can be determined.

$$[Y(\omega)][V_{node}] = Y_s(\omega)[V_s] + [-I_{NL}] \qquad (7.1)$$

where:

- $[Y(w)]$ is the KxK admittance matrix of the associated linear network that results from nodal analysis.
- $[V_{node}]$ is a $Kx1$ vector consisting of the K node voltages.
- $Y_s(\omega)$ is the input source admittance.
- $[V_s]$ is a $Kx1$ vector consisting of all the independent voltage excitations.
- $[I_{NL}]$ is a $Kx1$ vector consisting of all the nonlinear currents entering or exiting each independent node in the system. Note that I_{NL} does not refer to the nonlinear currents entering at the nonlinear ports.

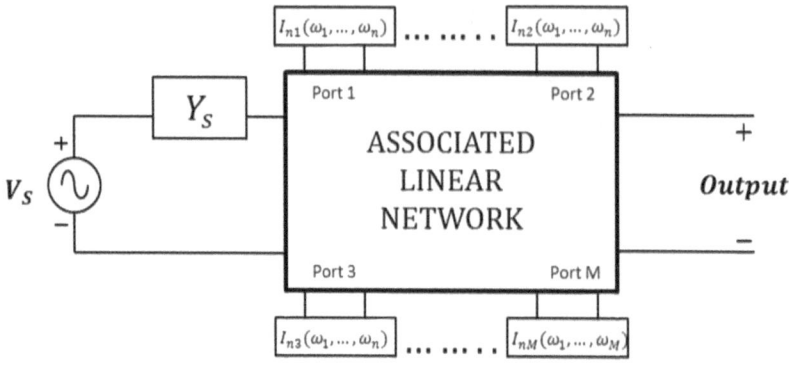

Figure 7.1c Associated linear network driven by
M nonlinear current sources

7. To solve for the first-order Volterra transfer function, $H_1(\omega_1)$, the input is driven by $e^{j\omega_1 t}$ and a harmonic balance is performed for the harmonics at ω_1. Recall that the frequency-domain representations of the nonlinear currents are the coefficients of harmonics at linearly independent frequencies (i.e. $\omega_1 + \omega_2, \omega_1 + \omega_2 + \omega_3, \dots$). Therefore, they do not contribute to the first-order transfer function and can be ignored in (7.1). The result is (7.2). This step is equivalent to open-circuiting all the nonlinear current sources and solving the resulting associated linear network for $[H_1(\omega_1)]$, the first-order Volterra transfer function vector (see Figure7.1d).

$$[Y(\omega_1)][H_{1node}(\omega_1)] = Y_s(\omega_1) \cdot \begin{bmatrix} 1 \\ 0 \\ 0 \\ \cdot \\ \cdot \end{bmatrix} \qquad (7.2)$$

where:

$$[H_{1node}(\omega_1)] = \begin{bmatrix} H_{1a}(\omega_1) \\ H_{1b}(\omega_1) \\ H_{1c}(\omega_1 \\ . \\ . \\ . \\ H_{1K}(\omega_1) \end{bmatrix}$$

- is a $Kx1$ vector representing the input-to-node first-order transfer functions. The subscripts $a, b, c, ..., K$ represent the independent nodes of the associated linear network.

- $[Y(\omega_1)]$ is the KxK admittance matrix representing the associated linear network evaluated at ω_1.

- $Ys(\omega_1)$ is the input source admittance evaluated at ω_1.

- $\begin{bmatrix} 1 \\ 0 \\ 0 \\ . \\ . \end{bmatrix}$

  is a $Kx1$ vector representing the independent voltage excitations. The 1 in the first row comes from the fact that when a linear network is driven by the harmonic $e^{j\omega_1 t}$ the response is the first-order transfer function evaluated at ω_1 times the same harmonic. Therefore, the term $e^{j\omega_1 t}$ can be divided from both sides of the system equation. The excitation is located in the first row because it drives the first independent node of the associated linear network.

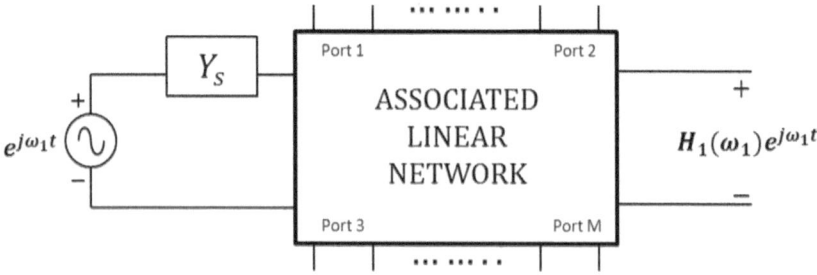

Figure 7.1d First-order Solution

Solving (7.2) for the first-order Volterra transfer function vector yields:

$$
\begin{bmatrix} H_{1a}(\omega_1) \\ H_{1b}(\omega_1) \\ H_{1c}(\omega_1 \\ \cdot \\ \cdot \\ \cdot \\ H_{1K}(\omega_1) \end{bmatrix} = [Y(\omega_1)]^{-1} \cdot Y_s(\omega_1) \cdot \begin{bmatrix} 1 \\ 0 \\ 0 \\ \cdot \\ \cdot \end{bmatrix} \tag{7.3}
$$

8. To solve for the second-order Volterra transfer function the analysis again begins with (7.1). This time however, the harmonics are balanced at $\omega_1 + \omega_2$. Recall that this is accomplished by driving the input with the harmonic sum $e^{j\omega_1 t} + e^{j\omega_2 t}$. Therefore, when balancing the harmonics at $\omega_1 + \omega_2$, the $Kx1$ vector of independent voltage excitations can be ignored. The result is (7.4). This is equivalent to shorting the input voltage source, Vs, and driving the associated linear network with the frequency-

domain representation of the second-order nonlinear currents (see Figure 7.1e).

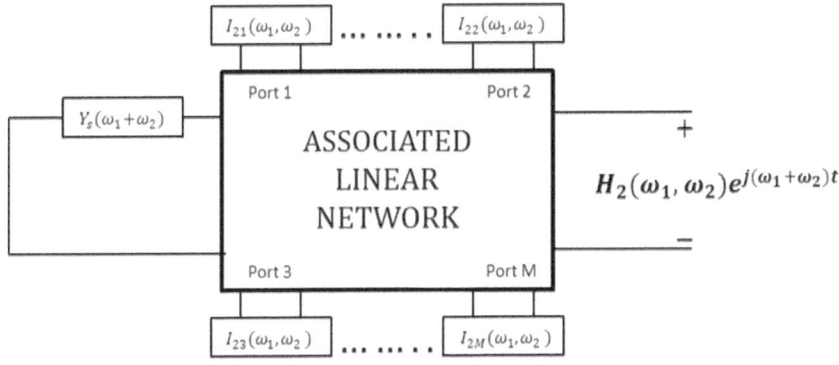

Figure 7.1e Second-order Solution

$$[Y(\omega_1 + \omega_2)][H_{2node}(\omega_1, \omega_2)] = [-I_2(\omega_1, \omega_2)] \qquad (7.4)$$

where:

- $$[H_{2node}(\omega_1, \omega_2)] = \begin{bmatrix} H_{2a}(\omega_1, \omega_2) \\ H_{2b}(\omega_1, \omega_2) \\ H_{2c}(\omega_1, \omega_2) \\ \cdot \\ \cdot \\ \cdot \\ H_{2K}(\omega_1, \omega_2) \end{bmatrix}$$

is a $Kx1$ vector representing the input-to-node second-order transfer functions. The subscripts $a, b, c, \ldots, K$ again refer to the independent nodes of the associated linear network.

- $[Y(\omega_1 + \omega_2)]$ is the KxK admittance matrix representing the associated linear network evaluated at $\omega_1 + \omega_2$.

- $Y_s(\omega_1 + \omega_2)$ is the input source admittance evaluated at $\omega_1 + \omega_2$.

- $[-I_2(\omega_1, \omega_2)] = \begin{bmatrix} -I_{2a}(\omega_1, \omega_2) \\ -I_{2b}(\omega_1, \omega_2) \\ -I_{2c}(\omega_1, \omega_2) \\ \cdot \\ \cdot \\ \cdot \\ -I_{2K}(\omega_1, \omega_2) \end{bmatrix}$

  are the frequency-domain second-order nonlinear currents flowing into or out of the independent nodes $a, b, c, \dots, K$.

Solving (7.4) for the second-order Volterra transfer function vector yields:

$$\begin{bmatrix} H_{2a}(\omega_1, \omega_2) \\ H_{2b}(\omega_1, \omega_2) \\ H_{2c}(\omega_1, \omega_2) \\ \cdot \\ \cdot \\ \cdot \\ H_{2K}(\omega_1, \omega_2) \end{bmatrix} = [Y(\omega_1 + \omega_2)]^{-1} \cdot \begin{bmatrix} -I_{2a}(\omega_1, \omega_2) \\ -I_{2b}(\omega_1, \omega_2) \\ -I_{2c}(\omega_1, \omega_2) \\ \cdot \\ \cdot \\ \cdot \\ -I_{2K}(\omega_1, \omega_2) \end{bmatrix} \qquad (7.5)$$

9. Continue this process until the desired order Volterra transfer function has been determined (see Figure 7.1f).

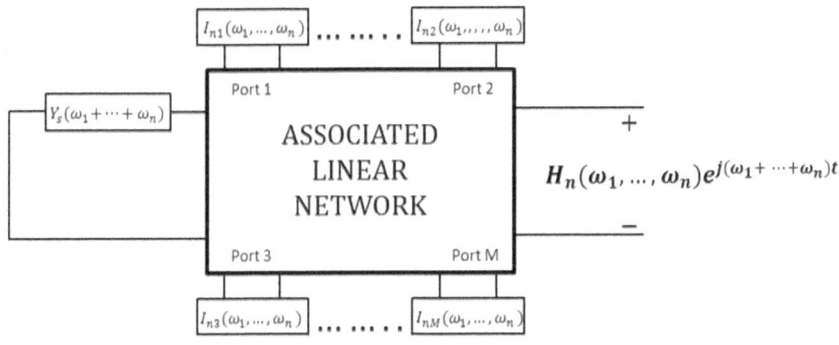

Figure 7.1f nth-order Solution

$$[H_{nnode}(\omega_1, \dots, \omega_n)] = [Y(\omega_1 + \dots + \omega_n)]^{-1} \begin{bmatrix} -I_{na}(\omega_1, \dots, \omega_n) \\ -I_{nb}(\omega_1, \dots, \omega_n) \\ -I_{nc}(\omega_1, \dots, \omega_n) \\ \cdot \\ \cdot \\ \cdot \\ -I_{nK}(\omega_1, \dots, \omega_n) \end{bmatrix}$$

$$(7.6)$$

7.3 APPLICATION

As an example, the MESFET circuit introduced in Chapter 3.0 will be analyzed using this method.

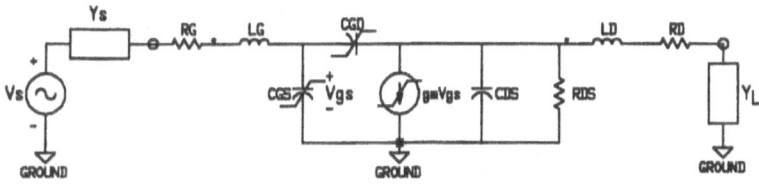

Figure 7.2 Complete MESFET Model

In this case the dominant nonlinearities are assumed to be the reverse-biased Schottky diode modeled by C_{GS}, and the transconductance g_m. C_{GD} also represents a reverse-biased Schottky diode, however under typical bias conditions (i.e. $V_{GD} > 3$ volts) its nonlinear behavior is very weak and therefore, can be modeled by its linearized value.

Note the linearization of C_{GD} is motivated by the desire to simplify the analysis and not due to the inability of Volterra series analysis to handle a nonlinearity between these two nodes. When V_{GD} is much less than 3 volts its true nonlinear nature should be incorporated into the analysis. Figure 7.2 can be simplified as shown in Figure 7.3.

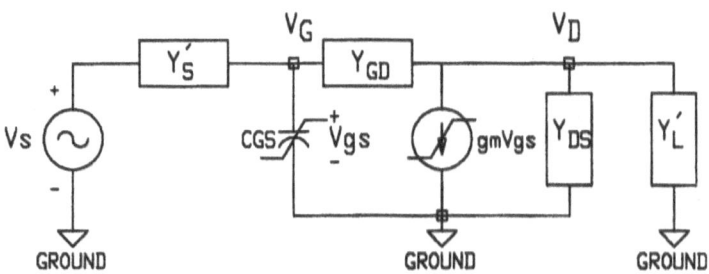

Figure 7.3 Simplified MESFET Model

where:

- $Y_S' = (Z_s + R_G + j\omega L_G)^{-1}$
- $Y_{GD} = j\omega C_{GD}$
- $Y_{DS} = G_{DS} + j\omega C_{DS}$
- $Y_L' = (Z_L + R_D + j\omega L_D)^{-1}$

In this way, the number of independent nodes is cut in half.

APPLYING STEPS 1 & 2:

The transconductance and the input gate-to-source channel capacitance are principally functions of the gate voltage, which controls the depletion depth in the channel. Their voltage dependence can be modeled by a power series of the form:

$$g_m(v_G) = g_{m1} + g_{m2}v_G + g_{m3}v_G^2$$

$$g_m(v_G) = \sum_{k=1}^{3} g_{mk} v_G^{k-1} \qquad (7.7)$$

$$C_{GS}(v_G) = C_{GS1} + C_{GS2}v_G + C_{GS3}v_G^2$$

$$C_{GS}(v_G) = \sum_{k=1}^{3} C_{GSk} v_G^{k-1} \qquad (7.8)$$

APPLYING STEPS 3 & 4:

Because the forms of both nonlinearities are identical to the ones analyzed in Chapter 6, the resulting equations (6.26), (6.28), (6.32) and (6.34) can be used to express the frequency-domain nonlinear currents for this example.

Nonlinear Transconductance:

$$I_{2D}(\omega_1, \omega_2) = g_{m2}H_{1G}(\omega_1)H_{1G}(\omega_2) \qquad (7.9)$$

$$I_{3D}(\omega_1, \omega_2, \omega_2) = \frac{2}{3}g_{m2}\begin{bmatrix} H_{1G}(\omega_1)H_{2G}(\omega_2, \omega_3) \\ + H_{1G}(\omega_2)H_{2G}(\omega_1, \omega_3) \\ + H_{1G}(\omega_3)H_{2G}(\omega_1, \omega_2) \end{bmatrix}$$

$$+ g_{m3}H_{1G}(\omega_1)H_{1G}(\omega_2)H_{1G}(\omega_3) \qquad (7.10)$$

Note the similarity of (7.9) and (7.10) to (6.26) and (6.28) respectively.

Nonlinear Capacitance:

$$I_{2G}(\omega_1, \omega_2) = C_{GS2}j(\omega_1 + \omega_2)H_{1G}(\omega_1)H_{1G}(\omega_2) \qquad (7.11)$$

$$I_{3G}(\omega_1, \omega_2, \omega_2) = \frac{2}{3}C_{GS2}j(\omega_1 + \omega_2 + \omega_2)\begin{bmatrix} H_{1G}(\omega_1)H_{2G}(\omega_2, \omega_3) \\ + H_{1G}(\omega_2)H_{2G}(\omega_1, \omega_3) \\ + H_{1G}(\omega_3)H_{2G}(\omega_1, \omega_2) \end{bmatrix}$$
$$+ j(\omega_1 + \omega_2 + \omega_2)C_{GS3}H_{1G}(\omega_1)H_{1G}(\omega_2)H_{1G}(\omega_3)$$
$$\qquad (7.12)$$

Again, observe the similarity of (7.11) and (7.12) to (6.32) and (6.34) respectively.

$H_{1G}(\cdot)$ and $H_{2G}(\cdot)$ represent the first- and second-order Volterra transfer functions at node G.

APPLYING STEP 5:

Replacing the nonlinear elements with their frequency-domain Norton equivalents yields the circuit shown in Figure 7.4.

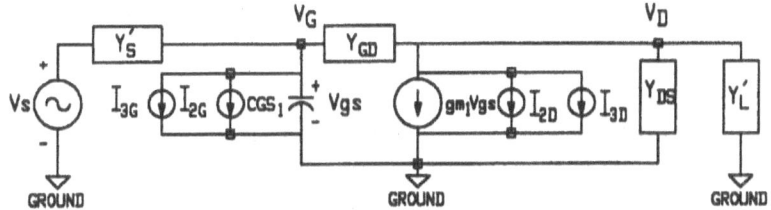

Figure 7.4 MESFET Model with Nonlinear Currents

APPLYING STEP 6:

Using nodal analysis determine the system equations.

$$[Y(\omega)][V] = Y_S'(\omega)[V_s] + [-I_{NL}]\tag{7.13}$$

$$\begin{bmatrix} Y_S' + Y_{GD} - j\omega_1 C_{GS1} & -Y_{GD} \\ Y_{GD} + g_{m1} & Y_{DS} + Y_L' - Y_{GD} \end{bmatrix}\begin{bmatrix} V_G \\ V_D \end{bmatrix} = Y_S'\begin{bmatrix} V_S \\ 0 \end{bmatrix} + \begin{bmatrix} -I_{NLG} \\ -I_{NLD} \end{bmatrix}\tag{7.14}$$

APPLYING STEP 7:

Open the nonlinear current sources and solve for $[H_1(\omega_1)]$ by

letting $V_S = 1 \Rightarrow V_G = H_{1G}(\omega_1)$

$$V_D = H_{1D}(\omega_1)$$

$$\begin{bmatrix} H_{1G}(\omega_1) \\ H_{1D}(\omega_1) \end{bmatrix} = \begin{bmatrix} Y_S' + Y_{GD} - j\omega_1 C_{GS1} & -Y_{GD} \\ Y_{GD} + g_{m1} & Y_{DS} + Y_L' - Y_{GD} \end{bmatrix}^{-1} \cdot Y_S'(\omega_1)\begin{bmatrix} 1 \\ 0 \end{bmatrix}\tag{7.15}$$

APPLYING STEP 8:

Short circuit the input voltage source and connect the second-order nonlinear currents to their respective nonlinear ports. Solve for $[H_2(\omega_1, \omega_2)]$.

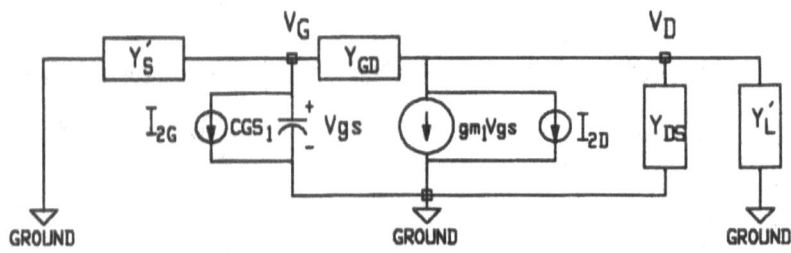

Figure 7.5 MESFET Model with Second-order Nonlinear Currents

$$[Y(\omega_1 + \omega_2)][H_2(\omega_1, \omega_2)] = [-I_2(\omega_1, \omega_2)] \qquad (7.16)$$

$$[H_2(\omega_1, \omega_2)] = [Y(\omega_1 + \omega_2)]^{-1} \cdot [-I_2(\omega_1, \omega_2)] \qquad (7.17)$$

$$\begin{bmatrix} H_{2G}(\omega_1, \omega_2) \\ H_{2D}(\omega_1, \omega_2) \end{bmatrix} = [Y(\omega_1 + \omega_2)]^{-1} \cdot \begin{bmatrix} -I_{2G}(\omega_1, \omega_2) \\ -I_{2D}(\omega_1, \omega_2) \end{bmatrix} \qquad (7.18)$$

$$\begin{bmatrix} H_{2G}(\omega_1, \omega_2) \\ H_{2D}(\omega_1, \omega_2) \end{bmatrix} =$$

$$[Y(\omega_1 + \omega_2)]^{-1} \cdot \begin{bmatrix} -C_{GS2}j(\omega_1 + \omega_2)H_{1G}(\omega_1)H_{1G}(\omega_2) \\ -g_{m2}H_{1G}(\omega_1)H_{1G}(\omega_2) \end{bmatrix} \qquad (7.19)$$

APPLYING STEP 9:

Applying the same process to the third-order case yields:

$$\begin{bmatrix} H_{3G}(\omega_1,\omega_2,\omega_3) \\ H_{3D}(\omega_1,\omega_2,\omega_3) \end{bmatrix} = [Y(\omega_1+\omega_2+\omega_3)]^{-1} \cdot \begin{bmatrix} -I_{3G}(\omega_1,\omega_2,\omega_3) \\ -I_{3D}(\omega_1,\omega_2,\omega_3) \end{bmatrix}$$

(7.20)

$$\begin{bmatrix} H_{3G}(\omega_1,\omega_2,\omega_3) \\ H_{3D}(\omega_1,\omega_2,\omega_3) \end{bmatrix} = [Y(\omega_1+\omega_2+\omega_3)]^{-1}$$

$$\cdot \begin{bmatrix} -\frac{2}{3}C_{GS2}j(\omega_1+\omega_2+\omega_3)\begin{bmatrix} H_{1G}(\omega_1)H_{2G}(\omega_2,\omega_3) \\ + H_{1G}(\omega_2)H_{2G}(\omega_1,\omega_3) \\ + H_{1G}(\omega_3)H_{2G}(\omega_1,\omega_2) \end{bmatrix} \\ +j(\omega_1+\omega_2+\omega_3)C_{GS3}H_{1G}(\omega_1)H_{1G}(\omega_2)H_{1G}(\omega_3) \\ \hline -\frac{2}{3}g_{m2}\begin{bmatrix} H_{1G}(\omega_1)H_{2G}(\omega_2,\omega_3) \\ + H_{1G}(\omega_2)H_{2G}(\omega_1,\omega_3) \\ + H_{1G}(\omega_3)H_{2G}(\omega_1,\omega_2) \end{bmatrix} + g_{m3}H_{1G}(\omega_1)H_{1G}(\omega_2)H_{1G}(\omega_3) \end{bmatrix}$$

(7.21)

Neither $H_{nG}(\cdot)$ nor $H_{nD}(\cdot)$ represent the transfer function from input-to-output due to the circuit simplification done prior to STEPS 1 & 2. $H_{nL}(\cdot)$ can be determined by applying voltage division to $H_{nD}(\cdot)$.

$$H_{1L}(\omega_1) = \left[\frac{Z_L(\omega_1)}{Z_L(\omega_1)+R_D+j\omega_1 L_D}\right] \cdot H_{1D}(\omega_1)$$

(7.22)

$$H_{2L}(\omega_1,\omega_2) = \left[\frac{Z_L(\omega_1+\omega_2)}{Z_L(\omega_1+\omega_2)+R_D+j(\omega_1+\omega_2)L_D}\right] \cdot H_{2D}(\omega_1,\omega_2)$$

(7.23)

$$H_{3L}(\omega_1,\omega_2,\omega_3) =$$

$$\left[\frac{Z_L(\omega_1+\omega_2+\omega_3)}{Z_L(\omega_1+\omega_2+\omega_3)+R_D+j(\omega_1+\omega_2+\omega_3)L_D}\right] \cdot H_{3D}(\omega_1,\omega_2,\omega_3)$$

(7.24)

This example demonstrates that combining the method of nonlinear currents, harmonic input method and linear system theory yields a systematic, straightforward, technique for solving multi-node nonlinear networks. This technique involves the solution of a system of linear differential equations driven by both linear and nonlinear sources.

7.4 CONVERGENCE REVISITED

This is an appropriate time to reexamine the issue of convergence. Little has been said prior to this because determining whether a Volterra series converged for a particular circuit was too difficult. It meant solving the d'Alambert criterion as described in Section 4.1.

$$\frac{1}{R} = lim_{n \to \infty} \left| \frac{g_{n+1}}{g_n} \right|$$

where $g_n = \int_{-\infty}^{\infty} \dots \int_{-\infty}^{\infty} |h_n(\tau_1, \tau_2, \dots \tau_n)| d\tau_1 d\tau_2 \dots d\tau_n$

For all practical purposes, this procedure is an extremely difficult task. First, to apply the d'Alambert criterion, the Volterra kernel and not the transfer function is required. Therefore, an inverse Fourier transform must be performed if only the transfer function is known. Second, the nth-order kernel needs to be determined. This requires that enough lower-order kernels be calculated so that a general trend, and therefore, a general expression, can be established. And finally, a multidimensional integral must be solved. This process is impractical and results in the question of convergence being assumed or totally ignored.

From work in previous sections, a simple easy-to-determine criterion with two conditions can be established, which guarantees the existence and convergence of a Volterra series for a particular nonlinear system [3].

1. The nonlinearities must be analytic in the region of interest.
2. The linearized network must be asymptotically stable in the region of interest.

The first condition states that the nonlinear system should be operated in a region where the nonlinear constitutive relationships are expandable in power series with nonzero radii of convergence, in other words, operate in a region where the constitutive relationships are analytic. This is not an overly strict condition. The entire formulation of Volterra series analysis has been based on this fact. This is another way of stating that Volterra series analysis is only valid for weakly nonlinear systems.

The second condition is also not very restrictive. It establishes the requirement that the first-order impulse response tends to zero as time approaches infinity:

$$h_1(t) \xrightarrow[as\ t\to\infty]{} 0$$

When this condition is met, the system can be shown to be "BIBO" stable. If this condition is not met, the system will not have a well-defined steady-state behavior. This condition was only placed on the first-order impulse response and not higher-order ones. The need to apply this constraint only to the first-order impulse response results from the previous observations that all higher-order transfer functions, and through inverse Fourier transforms their kernels, can be represented entirely in terms of $H_1(\omega_1)$ and $h_1(\tau_1)$ respectively. The importance of the linearized system was particularly highlighted in the chapter on the method of nonlinear currents.

Both these conditions make good engineering sense, are easy to establish and apply to many circuits encountered in electrical engineering.

8.0 QUANTIFICATION OF NONLINEAR BEHAVIOR

8.1 INTRODUCTION

Now that an analysis methodology for computing the different orders of Volterra transfer functions has been established, the next logical step is to determine how they can be used to quantify a system's nonlinear behavior.

A number of different measurement and analysis schemes of varying complexity exist to accomplish this task. Some of the more common methods, together with a brief description, are listed below:

8.2 HARMONIC DISTORTION

Harmonic distortion [8], [28] is a measure of the relative power of the harmonics generated by a nonlinear system. The most common way to quantify this phenomenon in the microwave world is dB relative to the fundamental or, thinking about a communication system, the "carrier". This is referred to as dBc. If a harmonic has 20 dB less power than the fundamental, then it is said to be -20 dBc:

$$dBc_{Harmonic} = 10 \log_{10} \left[\frac{Harmonic\ Power}{Fundamental\ Power} \right] \qquad (8.1)$$

Another way to quantify the harmonic distortion is with the term total-harmonic-distortion, THD. THD is the ratio of the root-mean-square voltage due to all the harmonics to the magnitude of the fundamental voltage:

$$THD = \frac{[|V_2|^2+|V_3|^2+|V_4|^2+\cdots]^{\frac{1}{2}}}{|V_1|} \tag{8.2}$$

8.3 INTERMODULATION DISTORTION

Intermodulation distortion [8], [11], [25] is a measure of the frequency translation nature of the nonlinear system. Typically this is the most detrimental type of distortion in many nonlinear systems because of the difficulty in filtering out some of the mixing products generated. Of greatest concern are the second- and third-order intermodulation distortion (IMD) products. In particular, IM_3 generates $2\omega_2 - \omega_1$ and $2\omega_1 - \omega_2$ which can be in the passband. The measure of either one requires a two-tone test. In this test two excitation signals of equal magnitude and very close in frequency are applied to the input of the nonlinear circuit. The resultant mixing products are then measured. Typical figures of merit are called the intermodulation intercept points (IP_n). These define theoretical points where the output power of the fundamental and the output power of the nth-order mixing product intersect. These are only theoretical quantities because both the first-order and IM products eventually saturate due to the system's finite power capability (i.e. power supply limits).

If the two input signals are at ω_1 and ω_2 respectively, then the second-order intermodulation term is defined by the power in the $\omega_2 - \omega_1$ or $\omega_1 - \omega_2$ mixing product and its intercept point is referred to as IP_2. The third-order case is defined by the power in the $2\omega_2 - \omega_1$ or $2\omega_1 - \omega_2$ mixing product and its intercept point is called IP_3.

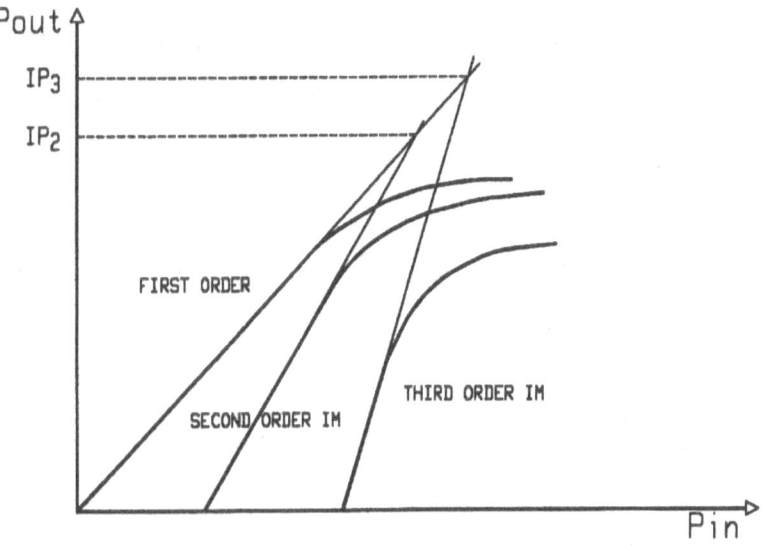

Figure 8.1 Plot of 2nd- and 3rd-order Intercept Points.

8.4 GAIN COMPRESSION

Gain compression [8], [25], [28] is the last method to be presented and the one that will be explored in greatest detail.

In most applications the third-order term is the one of most interest because the IM_3 products are particularly hard or impossible to filter out as mentioned earlier, and because they determine in large measure the manner in which a system saturates.

A relatively easy and accurate method for quantifying gain compression is to measure the input or output power at which the gain at the fundamental frequency decreases by 1 dB from its linear value.

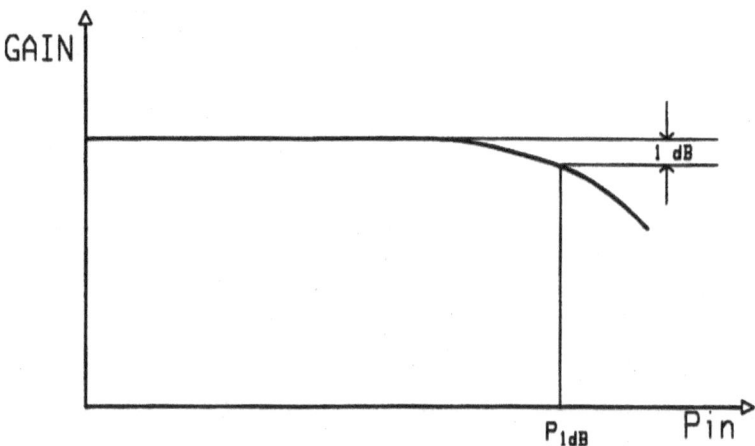

Figure 8.2 Plot of Gain Compression.

To understand how gain compression comes about, Equation (4.76) is required. This equation represents the output of a nonlinear system to the sinusoidal input, $x(t) = A \cos(\omega_0 t)$.

132

$$y(t) = \sum_{n=1}^{N} \frac{A^n}{2^n} \sum_{k=0}^{n} \binom{n}{k} H_{k,n-k}(\omega_0) e^{j(n-2k)\omega_0 t} \qquad (4.76)$$

Expanding (4.76) for $N = 3$ and collecting all the terms at the fundamental frequency ω_0 yields:

$$\begin{aligned}
y(t)|_{@\omega_0} &= AH_1(\omega_0)e^{j\omega_0 t} + AH_1(-\omega_0)e^{-j\omega_0 t} \\
&\quad + \frac{3}{8}A^3 H_3(-\omega_0, \omega_0, \omega_0)e^{j\omega_0 t} \\
&\quad + \frac{3}{8}A^3 H_3(-\omega_0, -\omega_0, \omega_0)e^{-j\omega_0 t} \qquad (8.3)
\end{aligned}$$

Using the relationship $H_1^*(\omega_0) = H_1(-\omega_0)$, (8.3) can be simplified.

$$y(t)|_{@\omega_0} = \left[AH_1(\omega_0) + \frac{3}{4}A^3 H_3(-\omega_0, \omega_0, \omega_0) \right] \cos(\omega_0 t) \qquad (8.4)$$

Therefore, the expression for gain can be rewritten as:

$$Gain = H_1(\omega_0) + \frac{3}{4}A^2 H_3(-\omega_0, \omega_0, \omega_0) \qquad (8.5)$$

Equation (8.5) illustrates that the gain is a strong function of the input magnitude A. For small values of A the gain is dominated by the linear term H_1. But, as A increases, the third-order term H_3 begins to dominate. Another feature of (8.5) is the significance of the relative phases of H_1 and H_3. If $phase(H_1) \approx phase(H_3)$ then the gain increases with increasing input amplitude. This is known as gain enhancement. When $phase(H_1) \approx -phase(H_3)$ the gain decreases with increasing input amplitude and the more common phenomenon of gain saturation occurs. To quantify this characteristic, (8.5) is divided by $H_1(\omega_0)$ and set equal to $10^{-1/20}$.

$$20 \log_{10} \left| \frac{Gain}{H_1} \right| = -1 \text{ dB} \tag{8.6}$$

$$\left| \frac{Gain}{H_1} \right| = 10^{-1/20} \tag{8.7}$$

$$\left| 1 + \frac{3}{4} A^2 \frac{H_3(-\omega_0,\omega_0,\omega_0)}{H_1(\omega_0)} \right| = 0.89 \tag{8.8}$$

Manipulating (8.8) further yields:

$$A^4 \left[\frac{H_3(-\omega_0,\omega_0,\omega_0)}{H_1(\omega_0)} \right]^2 + 2.67 A^2 \frac{H_3(-\omega_0,\omega_0,\omega_0)}{H_1(\omega_0)} + 0.37 = 0 \tag{8.9}$$

Equation (8.9) can now be solved for the magnitude of the input signal, A. This value of A represents the amplitude of the input signal at which the gain is reduced by 1 dB and is called the '1 dB gain compression point'. This measure provides a quick method for comparing nonlinear systems.

An obvious advantage of this specification over intermodulation is the need for only one source. A less

134

obvious advantage has to do with the 1 dB gain compression point's independence from the load impedance. Typically, in high-frequency applications the output power is much more easily measured than the output voltage. From the measured power and knowledge of the load impedance, the output voltage can be calculated. In the gain compression method the entire measurement occurs at a single frequency and at one port, therefore, the load impedance cancels out. For harmonic distortion the impedance of the load at the fundamental and all relevant harmonics must be accurately known. For intermodulation distortion the load impedance at the desired mixing products must be known as well. Generally, accurate knowledge of the load's frequency response is not easy to determine. Therefore, whenever possible, methods that do not require accurate load impedance information should be employed.

9.0 CONCLUSION

This thesis has shown that a Volterra series is a useful tool for analyzing frequency-dependent analytic nonlinearities. The advantages of the Volterra series approach include:

- Ability to model frequency-dependent distortion.

- Ability to represent nonlinear behavior entirely in terms of circuit elements. This allows a more intuitive understanding of the nonlinearities and consequently how to enhance or mitigate them.

As with any method there are disadvantages that include:

- Difficulty in handling gross nonlinearities. Even when the system can be modeled by a convergent Volterra series, if the rate of convergence is not rapid, the higher-order terms cannot be ignored. For orders greater than three the mathematics become prohibitively difficult.

- Errors due to the truncation of series expansions at various steps. Generally truncation occurs three times during the analysis process:
  1. During the development of the power series representation of the global nonlinearity.
  2. During the Taylor series expansion about a bias point.

3. And finally during the expansion of the nonlinear system in a Volterra series.

At each stage, in order to keep the mathematics tractable, some limit on the order of the series must be imposed. The milder the nonlinearity, the smaller the error introduced by truncation.

- Finite radii of convergence. This results in a Volterra series not existing for some systems and only in a very limited input region for others.

This last limitation can be improved to allow a greater range of input signals by expanding the nonlinear constitutive relationships in a series of complete orthogonal basis functions:

$$f(x) = \sum_{n=1}^{\infty} c_n \Phi_n(x) \tag{9.1}$$

where $\int_a^b \Phi_n(x)\Phi_m(x)dx = \alpha \quad for \; m = n \tag{9.2}$
$$\qquad\qquad\qquad\qquad\quad 0 \quad for \; m \neq n$$

Examples of orthogonal functions include; Bessel, Legendre, Chebycheff, Sinusoidal, etc.

Typically, the coefficient c_n is determined by minimizing the integral-square error I_N:

$$I_N = \int_a^b [f(x) - \sum_{n=1}^{N} c_n \Phi_n(x)]^2 dx \tag{9.3}$$

This integral will exist only if:

$$\int_a^b f^2(x)dx < \infty \tag{9.4}$$

By requiring that the basis functions form a complete set, (9.1) converges in the mean to any piecewise continuous function $f(\cdot)$ that satisfies (9.4) [22]. Therefore, over the range $a < x < b$ the convergence problems associated with Taylor series expansions have been eliminated.

An obvious question is why an orthogonal series is guaranteed to converge to a function $f(x)$ over a given range $[a, b]$ whereas a Taylor series is not. The answer lies in the constraints placed on the function $f(x)$. For a Taylor series approximation the function $f(x)$ must be analytic while, in an orthogonal series approximation, $f(x)$ need only be piecewise continuous. The result is that for a Taylor series approximation, not only must the difference between the function and the approximation tend to zero as $N \to \infty$, but also all the derivatives of the difference. For an orthogonal series approximation, only the area under the square of the error must approach zero as $N \to \infty$. It is this less restrictive form of convergence that facilitates a larger range of convergence for an orthogonal series expansion of the function, $f(x)$ [23].

REFERENCES

[1]   Arfken G.: "Mathematical Methods for Physicists",
      Academic Press Inc., San Diego, 1985., pp. 282.

[2]   Bedrosian E., Rice S.: "The Output Properties of
      Volterra Systems (Nonlinear Systems with Memory)
      Driven by Harmonic and Gaussian Inputs", Proceedings of
      the IEEE, vol. 59, No. 12., pp. 1688-1707, 1971.

[3]   Boyd S., Chua L.O.: "Volterra Series for Nonlinear
      Circuits", International Symposium on Circuits and
      Systems Proceedings, Vol. 1, 1985, pp. 369.

[4]   Boyd S., Chua L.O.: "Fading Memory and the Problem of
      Approximating Nonlinear Operators with Volterra
      Series", Trans. IEEE Circuits Syst., Vol. CAS-32, 1985,
      No. 11, pp. 1150-1161.

[5]   Boyd S.: "Volterra Series: Engineering Fundamentals",
      Ph.D dissertation, University of California, Berkeley,
      1985.

[6]   Bussgang, S. Ehrman, L. Graham, J.: "Analysis of
      Nonlinear Systems with Multiple Inputs", Proceedings of
      the IEEE, vol. 62, pp. 1088-1119, Aug. 1974.

[7]   Hung G., Stark L., Eykhoff P.: "On the Interpretation
      of Kernels", Annals of Biomedical Engineering, 1977,
      vol. 5, pp. 130-143.

[8]   Jerse T.: "Nonlinear Analysis of Microwave Circuits",
      EEC 289K Class notes, University of California,
      Davis, 1990.

[9]   Krozer V., Hartnagel H.: "Intermodulation Distortion
      Analysis of Cascaded MESFET Amplifiers using Volterra
      Series Representation", Int. J. Electronics, 1985,
      vol. 58, No. 4, pp. 693-708.

[10] Kundert K., Sangiovanni-Vicentelli S.: "Simulation of Nonlinear Circuits in the Frequency Domain", IEEE Transactions on Computer-Aided Design, vol. CAD-5, No. 4, pp. 521-535, Oct. 1986.

[11] Maas S.A.: "Nonlinear Microwave Circuits", Artech House, Norwood, MA, 1988.

[12] Minasian, R.: "Intermodulation Distortion of a MESFET Amplifier using the Volterra Series Representation", IEEE Trans. Microwave Theory Tech., MTT-28, pp. 1-8, Jan. 1980.

[13] Minasian, R.: "Simplified GaAs MESFET Model to 10 GHz", Electron. Lett., vol. 13, pp. 549-551, 1977.

[14] Minasian, R.: "Large Signal GaAs MESFET Model and Distortion Analysis", Electron. Lett., vol. 14, pp. 183-185, 1978.

[15] Narayanan, S.: "Application of Volterra Series to Intermodulation Distortion of a Transistor Feedback Amplifier", IEEE Trans. Circuit Theory, CT-17, pp. 518-527, Nov. 1970.

[16] Narayanan S.: "Transistor Distortion Analysis Using Volterra Series Representation", The Bell System Technical Journal, vol. 46, 1967, pp. 991-1023.

[17] Papoulis, A.: "The Fourier Integral and its Application", McGraw-Hill, 1962.

[18] Reiss W.: "A Convergence Condition of a Volterra Series for Harmonic Inputs", 1984 IEEE Int. Symp. Circuits & Systems Digest., 1984, pp. 563-6.

[19] Reiss W.: "Nonlinear Distortion Analysis of p-i-n Diode
Attenuators using Volterra Series Representation", 1984
IEEE Trans. Circuits & Systems, vol. CAS-31, No. 6,
June 1984, pp. 535-542.

[20] Reiss W.: "Volterra Series Representation of a Forward-
Biased p-i-n Diode", 1984 IEEE Trans. Electron Devices,
vol. ED-28, No. 12, Dec. 1981, pp. 1495-1500.

[21] Rugh, W.: "Nonlinear System Theory", The John Hopkins
University Press, Baltimore and London, 1981.

[22] Schetzen M.: "Nonlinear System Modeling Based on the
Wiener Theory", Proceedings of the IEEE, vol. 69,
No. 12., pp. 1557-1572, 1981.

[23] Schetzen, M.: "The Volterra and Wiener Theories of
Nonlinear Systems", John-Wiley & Sons, 1980.

[24] Schetzen M.: "Multilinear Theory of Nonlinear Systems",
Journal of the Franklin Institute, vol. 320, No. 5, pp. 221-
247, November 1985.

[25] Tri T. Ha: "Solid-State Microwave Amplifier Design",
John Wiley & Sons, 1981.

[26] Van den Eijnde E.: "Steady-State Analysis of Strongly
Nonlinear Circuits", Ph.D dissertation, Vrije
Universiteit Brussel, Brussel, 1989.

[27] Volterra, V.: "Theory of Functionals and of Integral
and Intro-differential Equations", Dover, New York, 1958.

[28] Weiner D., Spina J.: "Sinusoidal Analysis Modeling of
Weakly Nonlinear Circuits", Van Nostrand Reinhold, 1980.

[29] Wylie C.: "Advanced Engineering Mathematics",
McGraw-Hill, 1961.